# Azoles - Synthesis, Properties, Applications and Perspectives

*Edited by Aleksey Kuznetsov*

Published in London, United Kingdom

IntechOpen

*Supporting open minds since 2005*

Azoles - Synthesis, Properties, Applications and Perspectives
http://dx.doi.org/10.5772/intechopen.87765
Edited by Aleksey Kuznetsov

Contributors
Abdul Aziz Aziz Ali, Philippe Grellier, Roman Lesyk, Anna Kryshchyshyn, Danylo Kaminskyy, Seham A. Ibrahim, Hala F. Rizk, Brahim El Ibrahimi, Lei Guo, Nana Derkyi, Aleksey Kuznetsov



First published in London, United Kingdom, 2021 by IntechOpen
IntechOpen is the global imprint of INTECHOPEN LIMITED, registered in England and Wales, registration number: 11086078, 5 Princes Gate Court, London, SW7 2QJ, United Kingdom
Printed in Croatia

British Library Cataloguing-in-Publication Data
A catalogue record for this book is available from the British Library

Additional hard and PDF copies can be obtained from orders@intechopen.com

Azoles - Synthesis, Properties, Applications and Perspectives
Edited by Aleksey Kuznetsov
p. cm.
Print ISBN 978-1-83968-179-0
Online ISBN 978-1-83968-180-6
eBook (PDF) ISBN 978-1-83968-181-3

For EU product safety concerns:
IN TECH d.o.o., Prolaz Marije Krucifikse Kozulić 3, 51000 Rijeka, Croatia, info@intechopen.com or visit our website at intechopen.com.

# Meet the editor

Dr. Aleksey Kuznetsov pursued his Ph.D. in Physical Chemistry at the Department of Chemistry and Biochemistry, Utah State University, and graduated after three years of doctorate studies with a specialization in Computational/Theoretical Chemistry. He has been working in various subareas of this field of research since 2000. After several postdoctoral and visiting professor positions in Germany, the United States, and Brazil, Dr. Kuznetsov obtained a permanent faculty position at the Department of Chemistry, Universidad Técnica Federico Santa Maria, Santiago, Chile. He has been working there since 2019, focusing his research on the computational design of various complexes of porphyrins, including core-modified porphyrins, with nanoparticles, fullerenes, and graphenes, along with studies of transition metal complexes, organic compounds with pharmacological applications, and more.

# Contents

**Preface** **XI**

**Section 1**
Introduction **1**

**Chapter 1** **3**
Introductory Chapter: Azoles, Their Importance, and Applications
*by Aleksey E. Kuznetsov*

**Section 2**
Thiazoles and Their Derivatives: Synthesis and Applications **11**

**Chapter 2** **13**
Synthesis and Biological Evaluation of Thiazole Derivatives
*by Seham A. Ibrahim and Hala F. Rizk*

**Chapter 3** **33**
Thiazolidinone-Related Heterocyclic Compounds as Potential Antitrypanosomal Agents
*by Anna Kryshchyshyn, Danylo Kaminskyy, Philippe Grellier and Roman Lesyk*

**Section 3**
Triazoles: Synthesis and Applications **63**

**Chapter 4** **65**
1,2,3-Triazoles: Synthesis and Biological Application
*by Abdul Aziz Ali*

**Section 4**
Miscellaneous Applications of Azoles **83**

**Chapter 5** **85**
Azole-Based Compounds as Corrosion Inhibitors for Metallic Materials
*by Brahim El Ibrahimi and Lei Guo*

**Chapter 6** **113**
Azoles for Renewable Energy Development and Wood Treatment
*by Nana Derkyi*

# Preface

Azoles represent a broad, interesting, and promising class of five-membered heterocyclic aromatic compounds containing from one up to five nitrogen atom(s), which can also contain at least one sulfur or oxygen atom as a part of their conjugated ring (N,S and N,O subclasses of azoles, respectively). The parent azole compounds, as exemplified by imidazole, pyrazole, 1,2,3-triazole, tetrazole, and pentazole, are aromatic structures with two double bonds. Various successively reduced analogs, such as azolines and azolidines, with just one double bond, have been synthesized.

Only one lone pair of electrons from each heteroatom in the azole ring participates in the aromatic bonding. The numbering of ring atoms in azoles starts with the heteroatom that does not participate in the double bond and proceeds towards the other heteroatom. Imidazole, which contains two N atoms, and other five-membered aromatic heterocyclic compounds with two nitrogens (e.g., pyrazole) are extremely common in nature and form the core of many biomolecules, for instance, histidine, or parts of purine nucleobases.

Azoles have always been considered suitable scaffolding for the design of various novel therapeutic agents. Various oxygen-containing azoles, as exemplified by oxadiazoles, oxazoles, and isoxazoles, have been thoroughly studied for their diversified biological activities. Widely used as potent antifungal agents due to their valuable properties like a broad spectrum of action, chemical stability, and oral bioavailability, various azole derivatives have also demonstrated many other promising biological properties including antidiabetic, immunosuppressant, anti-inflammatory, and anticancer activities.

This book includes four sections. The first section contains an introductory chapter written by the editor and explains the importance and applications of Azoles. The second section, "Thiazoles and Their Derivatives: Synthesis and Applications," is composed of two chapters. Chapter 2, "Synthesis and Biological Evaluation of Thiazole Derivatives," presents the several types of thiazole-based heterocyclic scaffolds (monocyclic or bicyclic systems), their synthesis, studies of their biological activities, and the modifications of thiazole-based compounds to generate new molecules with potent antitumor, antioxidant, and antimicrobial activities. Chapter 3, "Thiazolidinone-Related Heterocyclic Compounds as Potential Antitrypanosomal Agents," describes the development of 4-thiazolidinone and thiazole derivatives with heterocyclic fragments, which exhibit good inhibition of trypanosome growth and might be potential candidates for the development of new drugs against trypanosomiasis.

The third section, "Triazoles: Synthesis and Applications," contains one chapter. Chapter 4, "1,2,3-Triazoles: Synthesis and Biological Application," considers numerous synthetic approaches for the synthesis of 1,2,3-triazoles, especially the popular click chemistry approach, and discusses several biological activities of these promising heterocycles.

The last section, "Miscellaneous Applications of Azoles," contains two chapters. Chapter 5, "Azole-Based Compounds as Corrosion Inhibitors for Metallic Materials," discusses the application of N-azole, N,S-azole (thiazole), and N,O-azole (oxazole) molecules and their derivatives as retarders of metallic corrosion as well as related highlighted outcomes in recent years. Chapter 6, "Azoles for Renewable Energy Development and Wood Treatment," provides a concise overview of integrating azoles in materials used for renewable energy processing and applications and wood treatment, with an outlook on challenges and opportunities.

**Aleksey Kuznetsov**
Department of Chemistry,
Universidad Técnica Federico Santa María,
Santiago, Chile

Section 1

# Introduction

Chapter 1

# Introductory Chapter: Azoles, Their Importance, and Applications

*Aleksey E. Kuznetsov*

## 1. Introduction

Heterocyclic compounds constitute an important and very broad class of organic molecules that are found to play a vital role in our daily life. The presence of the various heterocyclic frameworks in natural products and their widespread applications in the areas of material science, medicinal chemistry, agrochemicals, etc. emphasizes their extraordinary significance in diverse fields. Among the heterocyclic frameworks, structurally diverse azoles have been found to play an extremely significant role. Azoles represent a broad, very interesting, and perspective class of five-membered heterocyclic aromatic compounds whose framework contains from one and up to five nitrogen atom(s) and can also contain at least one S or O atom as a part of the azole conjugated ring (N,S and N,O subclasses of azoles, respectively) [1]. The parent azole compounds, as exemplified by imidazole, pyrazole, 1,2,3-triazole, tetrazole, and pentazole, are aromatic structures with two double bonds (**Figure 1**).

There have been synthesized various successively reduced analogs, such as azolines and azolidines, with just one double bond. Only one lone pair of electrons from each heteroatom in the azole ring participates in the aromatic bonding. The numbering of ring atoms in azoles starts with the heteroatom that does not participate in the double bond and proceeds toward the other heteroatom (**Figure 1**). Since the beginning of their studies and applications, major advances in the chemistry of pyrazoles, imidazoles, triazoles, tetrazoles, and their fused heterocyclic derivatives have been performed [2–6]. These azoles are also widely found as core structures in a large variety of natural and artificially synthesized compounds possessing important agrochemical and pharmaceutical properties [7–11]. The well-known ability of these heterocyclic cores to serve both as biomimetics and reactive pharmacophores encourages their applications in numerous drugs [12–17].

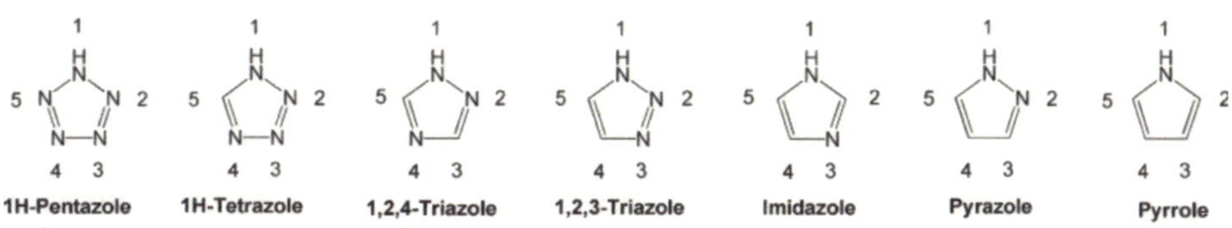


**Figure 1.**
*Structural formulae of unsubstituted neutral azoles (only includes nitrogen). Reprinted (adapted) with permission from ref. [2]. Copyright (2011) American Chemical Society.*

## 2. Coordination Chemistry of Azoles

Also, among other applications, which will be briefly touched in this chapter, azoles are known to play a significant role in coordination chemistry [18–22]. Thus, 1,2,3-triazoles (cf. **Figure 1**) have been known since the end of the 19th century, when 2-phenylbenzotriazole 1-oxide was described [23]. As early as 1937, the binding ability of the triazole ring was studied [24], although the first crystallographically characterized coordination compound was reported only in 1976 [25]. The isomeric to 1,2,3-triazoles 1,2,4-triazole ring was first mentioned in the end of the 19th century, in 1885 [18, 26]. Its ability to bind metal ions was established a few decades later [27] and the first crystal structure of one of the adducts was published already in 1962 [28]. In 1886, the term *tetrazole* was proposed for a five-membered heteroarene with four nitrogens [18], and in first decade of the 20th century, 1910, the potential binding of this heterocycle to metal ions was reported [18]. The first tetrazole complex characterized by X-ray diffraction study was reported in 1971 by Mason [10]. It should be noticed that although the chemistry of these three principal azoles has been studied for more than a century, their coordination behavior earlier was not the subject of extensive investigation [18, 27, 29]: structural reports of triazole- and tetrazole-based coordination compounds became increasingly common in the research literature only since the early 1980s [30]. The excellent 2011 review by Aromi et al. [18] clearly demonstrated the high versatility and suitability of the 1,2,3-triazole, 1,2,4-triazole, and tetrazole rings for the design and construction of outstanding coordination materials with attractive physicochemical properties. The straightforward preparation of such azole-containing ligands together with their synthetic flexibility allowed the syntheses of numerous outstanding systems such as coordination polymers and MOFs (metalloorganic frameworks), metal complexes, and coordination compounds with spin-crossover properties. It should be emphasized that these areas indeed represent current hot topics of investigation. Moreover, these N-donor ligands have found applications in many other fields of applied coordination chemistry, such as biological chemistry, nanomaterials, anion recognition, and nonlinear optics [18].

## 3. Azoles in Polymers

It is also of high interest to mention recent progress in azine- and azole-type N-heteroaromatic compounds for applications in structural engineering of high-mobility polymeric semiconductors [31]. The most fast developing area of polymeric semiconductors is production of novel semiconductors by employing the highly tunable donor–acceptor structural motifs. This approach revolutionized the whole strategy of the semiconducting polymers design. Furthermore, the appeal of replacing benzene or thiophene moieties with various $sp^2$-hybridized N-heteroaromatics, such as azine or azole heterocycles, directed design efforts toward developing materials with n-type or ambipolar charge transport behaviors. The nitrogen atoms introduced in polymer molecules allow to adjust molecular orbital energies, enhancing electron injection by lowering frontier molecular orbital energy levels. Moreover, they allow to reduce the steric effects which, in turn, results in maximizing electronic coupling. In this work an overview of recent progress in syntheses and characterization of azine- or azole-type N-heteroaromatics to be used in structural engineering of high-mobility polymeric semiconductors was given [31]. Various synthetic routes for creating these N-heteroaromatic building blocks and corresponding polymers were reviewed. These routes may inspire new developments in molecular engineering. Also, important structural features were discussed including the new semiconductor polymer electronic structures and conformational preferences. This review also discussed the correlations between the molecular structures of these N-heteroaromatic compounds and the device performances. To summarize, the semiconducting polymers containing N-heteroaromatic rings should be considered as primary candidates for functional design of compounds for specific applications in modern organic electronics.

Moreover, recently the latest achievements and problems associated with self-healing and shape memory metallopolymers (MP) such as metal complexes based on the polymers containing azole donor fragments among others (phenol, carboxylic acid, pyridine, histidine, and urethane) were reviewed [32]. Particular attention was paid to the principles of action of the shape memory MPs. MPs are in general of considerable interest due to their applications as functional materials for sensors, soft electronic devices, transistors, conductors, nanogenerators, bone tissue engineering, etc.

## 4. Azoles as Energetic Compounds

Of course, of a very high interest are applications of azoles and their derivatives as energetic compounds: thus, various azole-based energetic salts – tetrazole-based, triazole-based, imidazole and pyrazole-based – were reviewed by Gao and Shreeve in 2011 [2], and recently current synthesis and properties of energetic pentazolate and its derivatives were reviewed by Wozniak and Piercey [33]. The pentazolate, or cyclo-$N_5^-$, received increased attention in last two decades. Being the compound without carbons and hydrogens, the pentazolate anion is well known to release large amounts of energy upon decomposition simultaneously liberating environmentally friendly $N_2$ gas. Due to these extremely appealing qualities, the pentazolate anion and derivatives are essential in the development of novel high-energy-density materials. The review by Wozniak and Piercey considered the following aspects: (i) historical significance of cyclo-$N_5^-$; (ii) its precursors; (iii) synthesis routes of obtaining cyclo-$N_5^-$ with a focus on arylpentazole precursors; (iv) factors affecting the stability of cyclo-$N_5^-$; (v) energetic performances of currently used energetic cyclo-$N_5^-$-containing compounds; and (vi) future possible experimental research.

## 5. Azoles in Ionic Liquids

Furthermore, it is worthwhile to mention the review by Easton et al. [34] where azolate anions in ionic liquids (IL) were considered. Owing to their ease of synthesis, diffuse positive charge, and chemical stability, 1-alkyl-3-methylimidazolium cations are one of the most routinely utilized and historically important components in ionic liquid chemistry. However, the versatile chemistry of azoles to allow their use as an anionic component in ILs, as azolates, was investigated relatively scarcely. Azolate anions possess numerous desired properties for IL formation, such as diffuse ionic charge, tailorable asymmetry, and synthetic flexibility, with the added advantages of not relying on halogen atoms for electron withdrawing effects. The review explored the 122 known so far azolate-containing ionic liquids which were prepared from only 39 disparate azolate anions, with a goal to highlight not only their well pronounced utility as IL components, but also the ways in which their advantageous properties may be used by the broader scientific community for design of new tailored materials. In this context, it is also worthwhile to mention another work by Easton et al. [35] where a non-stoichiometric approach to control the solid-state behavior of protic ionic liquids (PILs) was demonstrated by direct mixing of 4,5-dicyanoimidazole (HDCNim) with either 1-ethylimidazole (C2im) or 1-butylimidazole (C4im) in different mole fractions.

## 6. Azoles as Corrosion Inhibitors

Also, azoles and their derivatives find numerous applications as organic corrosion inhibitors as was reviewed by Xhanari and Finšgar [36] and Fateh et al. [37]. In the first review, the authors summarized the research performed during the last two decades regarding the use of very important organic corrosion inhibitors for Al and its alloys in alkaline (mainly NaOH and KOH) and chloride solutions. The focus of this review was on the type of corrosion inhibitors and on their inhibition effectiveness and mechanism. The most frequently used corrosion inhibitors were shown to be the mercapto compounds, azole derivatives, organic dyes, and different polymers. Weight loss and electrochemical techniques were among the most frequently used techniques for evaluation of the corrosion inhibition effectiveness of the studied compounds. The second review covered corrosion of Cu and its alloys in corrosive environments along with their corrosion inhibitors. The main corrosion inhibitor groups for copper were introduced and a review of adsorption models was provided. The most widely used corrosion inhibitors for protection of copper in salt and weak acidic environments were shown to be organic compounds from azole family, such as triazole, benzotriazole, and thiazole, and for strong acidic media imidazol and tetrazole were demonstrated to perform the best. Also, it is worthwhile to mention the 2008 work by Kuznetsov and Kazansky [6].

## 7. Azoles in Chemosensors

Next, it is interesting to mention the recent developments in 1,2,3-triazole-based chemosensors as very recently reviewed by Ahmed and Xiong [38]. This review summarized the latest developments in the field of chemosensors based on click-generated triazoles which were used for detection of a range of metal cations, anions, and neutral analytes. The detection of metal ions became a significant and perspective field of research due to their medicinal, biological, and environmental impacts. This resulted in significant increase in the number of articles published on this subject, which reported more reliable and sophisticated triazole-based chemosensors for a variety of analytes. The review considered the development of chemosensors reported between 2012 and 2020 due to their advantages over other chemosensors, including such criteria as ease of recognition, simple instrumentation, along with high selectivity and high sensitivity.

## 8. Conclusions and Perspectives

Various oxygen-containing azoles, as exemplified by oxadiazoles, oxazoles, and isoxazoles, have been also thoroughly studied for their diversified biological activities. Widely used as potent antifungal agents (fungicides) due to their valuable properties like broad spectrum of action, chemical stability, and oral bioavailability [39–42], various azole derivatives have also demonstrated many other promising biological properties including antidiabetic, immunosuppressant, antiinflammatory, antiviral, antitubercular, and anticancer activities [8, 12, 42–46].

As can be seen, azoles have always been considered as an extremely suitable scaffold for the design of various novel therapeutic agents and other extremely versatile and useful compounds with potential applications in various areas such as materials, energetics, catalysis, etc. The intensive research work in the area of azoles, covering synthesis, characterization, and computational studies of their various novel derivatives is continuously ongoing [44, 47–52]. Thus, the area of azoles and their derivatives, their physico-chemical properties, and applications is of continuous high interest, and therefore this book will be a valuable addition to the knowledge which has been accumulated so far in this field.

## Author details

Aleksey E. Kuznetsov
Department of Chemistry, Universidad Técnica Federico Santa María, Santiago, Chile

*Address all correspondence to: aleksey.kuznetsov@usm.cl

## References

[1] Eicher, T.; Hauptmann, S.; Speicher, A. Five-Membered Heterocycles: Sections 5.22-5.36. In *The Chemistry of Heterocycles*; Wiley-VCH Verlag GmbH & Co. KGaA: Weinheim, Germany, 2013; pp. 122-184. ISBN 978-3-527-32747-8.

[2] Gao, H.; Shreeve, J. M. Azole-Based Energetic Salts. Chem. Rev., 2011, 111, 7377-7436.

[3] Moulin, A.; Bibian, M.; Blayo, A.-L.; El Habnouni, S.; Martinez, J.; Fehrentz, J.-A. Synthesis of 3,4,5-trisubstituted-1,2,4-triazoles. Chem. Rev., 2010, 110, 1809-1827.

[4] Frija, L. M. T.; Ismael, A.; Cristiano, M. L. S. Photochemical transformations of tetrazole derivatives: applications in organic synthesis. Molecules, 2010, 15, 3757-3774.

[5] Gomez-Zavaglia, A.; Reva, I. D.; Frija, L. M. T.; Cristiano, M. L. S.; Fausto, R. Molecular structure, vibrational spectra and photochemistry of 5-mercapto-1-methyltetrazole. J. Mol. Struc., 2006, 786, 182-192.

[6] Kuznetsov, Y. I.; Kazansky, L. P. Physicochemical aspects of metal protection by azoles as corrosion inhibitors. Russ. Chem. Rev., 2008, 77, 219-232.

[7] Fontana, G. Current Bbioactive azole-containing natural products. Curr. Bioact. Compd., 2010, 6, 284-308.

[8] Iradyan, M. A.; Iradyan, N. S.; Stepanyan, G. M.; Arsenyan, F. G.; Garibdzhanyan, B. T. Antitumor activity of imidazole derivatives: dacarbazine and the new alkylating agent imidazene (Review). Pharmaceut. Chem. J., 2010, 44, 175-182.

[9] Abele, E.; Abele, R.; Lukevics, E. Oximes of five-membered heterocyclic compounds with three and four heteroatoms 2. Synthesis of derivatives, reactions, and biological activity (Review). Chem. Heterocycl. Compd., 2008, 44, 769-792.

[10] Pibiri, I.; Buscemi, S. A recent portrait of bioactive triazoles. Curr. Bioact. Compd., 2010, 6, 208-242.

[11] Gaware, V. M.; Dighe, N. S.; Pattan, S. R.; Shinde, H. V.; Musmade, D. S.; Chavan, P. A.; Patel, P. Thiazolo-Triazole a nucleus possessing range of pharmacological activities: a review. Pharm. Lett., 2010, 2, 3540.

[12] Baroniya, S.; Anwer, Z.; Sharma, P. K.; Dudhe, R.; Kumar, N. Recent advancement in imidazole as anti-cancer agents: A review. Pharm. Sin. 2010, 1, 172-182.

[13] Myznikov, L. V.; Hrabalek, A.; Koldobskii, G. I. Drugs in the tetrazole series. Chem. Heterocycl. Compd., 2007, 43, 1.

[14] Zych, A. J.; Herr, R. J. Tetrazoles as carboxylic acid bioisosteres in drug discovery. Pharma. Chem., 2007, 6, 21-24.

[15] Kharb, R.; Sharma, Pr. C.; Yar, M. S. Pharmacological significance of triazole scaffold. J. Enzym. Inhib. Med. Chem., 2011, 26, 1-21.

[16] Nivoix, Y.; Ubeaud-Sequier, G.; Engel, P.; Leveque, D.; Herbrecht, R. Drug-drug interactions of triazole antifungal agents in multimorbid patients and implications for patient care. Curr. Drug Metab. 2009, 10, 395-409.

[17] Cronin, S.; Chandrasekar, P. H. Safety of triazole antifungal drugs in patients with cancer. J. Antimicrob. Chemother. 2010, 65, 410-416.

[18] Aromi, G.; Barrios, L. A.; Roubeau, O.; Gamez, P. Triazoles and tetrazoles: Prime ligands to generate remarkable coordination materials. Coord. Chem. Rev., 2011, 255, 485-546.

[19] McGinley, J.; Fleming, A. Synthesis of macrocycles containing tetrazole units—potential metal complexation sites. J. Inclusion Phenom. Macrocyclic Chem., 2008, 61, 1-10.

[20] Zhao, H.; Qu, Z.-R.; Ye, H.-Y.; Xiong, R.-G. *In situ* hydrothermal synthesis of tetrazole coordination polymers with interesting physical properties. Chem. Soc. Rev. 2008, 37, 84-100.

[21] Zhang, X.-M. Hydro(solvo)thermal *in situ* ligand syntheses. Coord. Chem. Rev. 2005, 249, 1201-1219.

[22] Pellei, M.; Lobbia, G. G.; Papini, G.; Santini, C. Synthesis and properties of poly(pyrazolyl)borate and related boron-centered scorpionate ligands. Part B: imidazole-, triazole- and other heterocycle-based systems. Mini-Rev. Org. Chem., 2010, 7, 173-203.

[23] Werner, A.; Stiasny, E. Ueber Nitroderivate des Azo-, Azoxy- und Hydrazo-Benzols. Berichte, 1899, 32. 3256-3282.

[24] Wilson, R. F.; Wilson, L. E. Preparation of palladium(II) chloride-1,2,3-benzotriazole coordination compounds. J. Am. Chem. Soc., 1955, 77, 6204-6205.

[25] Meunier-Piret, J.; Piret, P.; Putzeys, J. P.; Van Meerssche, M. Structure cristalline du complexe de l'hexakis-(benzotriazolyl)-hexakis(allylamine)-trisnickel(II) avec la triphénylphosphine oxide. Acta Crystallogr. B, 1976, 32, 714-717.

[26] Potts, K. T. The chemistry of 1,2,4-triazoles. Chem. Rev., 1961, 61, 87-127.

[27] Haasnoot, J. G. Mononuclear, oligonuclear and polynuclear metal coordination compounds with 1, 2, 4-triazole derivatives as ligands. Coord. Chem. Rev., 2000, 200, 131-185.

[28] Jarvis, J. A. J. The crystal structure of a complex of cupric chloride and 1:2:4-triazole. Acta Crystallogr., 1962, 15, 964-966.

[29] Fehlhamm, W.P.; Dahl, L. F. Structural characterization of a 1-substituted 5-tetrazolyl(transition metal) complex containing metal-carbon bonding. Stereochemistry of the tetrakis (1-isopropyltetrazol-5-ato) aurate(III) anion, $[Au(CN_4R)^{4-}$ (R = isopropyl). J. Am. Chem. Soc., 1972, 94, 3370-3377, and references therein.

[30] C. Temple Jr., *The Chemistry of Heterocyclic Compounds*, Wiley, New York, 1981.

[31] Huang, J.; Yu, G. Structural engineering in polymer semiconductors with aromatic N-heterocycles. Chem. Mater., 2021, 33, 1513 –1539.

[32] Dzhardimalieva, G. I.; Yadav, B. C.; Singh, S.; Uflyand, I. E. Self-healing and shape memory metallopolymers: state-of-the-art and future perspectives. Dalton Trans., 2020, 49, 3042-3087.

[33] Wozniak, D. R.; Piercey. D. G. Review of the current synthesis and properties of energetic pentazolate and derivatives thereof. Engin., 2020, 6, 981-991.

[34] Easton, M. E.; Choudhary, H.; Rogers., R. D. Azolate anions in ionic liquids: promising and under-utilized components of the ionic liquid toolbox. Chem. Eur. J., 2019, 25, 2127-2140

[35] Easton, M. E.; Li, K.; Titi, H. M.; Kelley, S. P.; Rogers., R. D. Controlling the interface between salts, solvates, co-crystals, and ionic liquids with non-stoichiometric protic azolium azolates. Cryst. Growth Des., 2020, 20, 2608–2616.

[36] Xhanari, K.; Finšgar, M. Organic corrosion inhibitors for aluminum and its alloys in chloride and alkaline solutions: A review. Arab. J. Chem., 2019, 12, 4646-4663.

[37] Fateh, A.; Aliofkhazraei, M.; Rezvanian, A. R. Review of corrosive environments for copper and its corrosion inhibitors. Arab. J. Chem., 2020, 13, 481-544.

[38] Ahmed, F.; Xiong, H. Recent developments in 1,2,3-triazole-based chemosensors. Dyes Pigm., 2021, 185, 108905.

[39] Campoy, S.; Adrio, J. L. Antifungals. Biochem. Pharm., 2017, 133, 86-96.

[40] Pérez-Cantero, A.; López-Fernández, L.; Guarro, J.; Capilla, J. Azole resistance mechanisms in Aspergillus: update and recent advances. Int. J. Antimicr. Agents, 2020, 55, 105807.

[41] Cai, W.; Ye, P.; Yang, B.; Shi, Z.; Xiong, Q.; Gao, F.; Liu, Y.; Zhao, J.; Ying. G. Biodegradation of typical azole fungicides in activated sludge under aerobic conditions. J. Env. Sci., 2021, 103, 288-297.

[42] Zhou, C.-H.; Wang, Y. Recent researches in triazole compounds as medicinal drugs. Curr. Med. Chem., 2012, 19, 239-280.

[43] Haroon, M.; Akhtar, T.; Khalid, M.; Ali, S.; Zahra, S.; ul haq, I.; Alhujaily, M.; de B. Dias, M. C. H.; Leite, A. C. L.; Muhammad, S. Synthesis, antioxidant, antimicrobial andantiviral docking studies of ethyl 2-(2-(arylidene) hydrazinyl)thiazole-4-carboxylates. Z. Naturforsch. C J. Biosci., 2021. Apr 26. doi: 10.1515/znc-2021-0042. Online ahead of print.

[44] Marín-Luna, M.; Claramunt, R. M.; Elguero, J.; Alkorta, I. Theoretical and spectroscopic characterization of API-related azoles in solution and in solid state. Curr. Pharm. Design, 2020, 26, 4847-4857.

[45] Ahmad, K.; Khan, M. K. A.; Baig, M. H.; Imran, M.; Gupta, G.K. Role of azoles in cancer prevention and treatment: present and future perspectives. Anti-Cancer Agents Med. Chem., 2018, 18, 46-56.

[46] Das, R.; Asthana, G. S.; Suri, K. A.; Mehta, D.; Asthana. A. Recent developments in azole compounds as antitubercular agent. Mini-Rev. Org. Chem., 2019, 16, 290-306.

[47] Mori, A. Structure- and functionality-based molecular design of azoles and thiophenes. Bull. Chem. Soc. Jpn., 2020, 93, 1200-1212.

[48] Lauko, J.; Kouwer, P. H. J.; Rowan, A. E. 1*H*-1,2,3-triazole: from structure to function and catalysis. J. Heterocyclic Chem., 2017, 54, 1677.

[49] De Oliveira Silva, A.; McQuade, J.; Szostak, M. Recent advances in the synthesis and reactivity of isothiazoles. Adv. Synthesis Cat., 2019, 361, 3050-3067.

[50] Zyryanov, G. V.; Kopchuk, D. S.; Kovalev, I. S.; Santra, S.; Rahman, M.; Khasanov, A. F.; Krinochkin, A. P.; Taniya, O. S.; Chupakhin, O. N.; Charushin, V. N. Rational synthetic methods in creating promising (hetero) aromatic molecules and materials. Mendeleev Commun., 2020, 30, 537-554.

[51] Reinus, B.; Kerwin, S. M. Preparation and utility of N-alkynyl azoles in synthesis. Molecules 2019, 24, 422.

[52] Narasimhamurthy, K. H.; Sajith, A. M.; Joy, M. N.; Rangappa, K. S. An overview of recent developments in the synthesis of substituted thiazoles. ChemistrySelect, 2020, 5, 5629-5656.

Section 2

# Thiazoles and Their Derivatives: Synthesis and Applications

Chapter 2

# Synthesis and Biological Evaluation of Thiazole Derivatives

*Seham A. Ibrahim and Hala F. Rizk*


## Abstract

Thiazoles belong to the group of azole heterocycles. They are aromatic five-membered heterocycles containing one sulfur and one nitrogen atom. In recent years thiazoles, their derivatives, and isomers have gained considerable attention because of their broad applications in different fields, such as agrochemicals, industrial, and photographic sensitizers. Also, they have pharmaceutical and biological activities that include antimicrobial (sulfazole), antiretroviral (ritonavir), antifungal (abafungin), anticancer (tiazofurin), antidiabetic, anti-inflammatory, anti-Alzheimer, antihypertensive, antioxidant, and hepatoprotective activities. The compounds containing thiazole moieties are a prominent structural feature in a variety of natural products, such as vitamin B and penicillin. Thus, in this chapter several types of thiazole-based heterocyclic scaffolds such as monocyclic or bicyclic systems synthesis and their biological activities studies are presented. Furthermore modification of thiazole-based compounds at different positions to generate new molecules with potent antitumor, antioxidant, and antimicrobial activities is described.


**Keywords:** azole heterocycles, thiazoles, biological activities, antioxidants, antimicrobial, anticancer, anti-Alzheimer, antihypertensive

## 1. Introduction

Thiazoles are five-membered heterocyclic compounds containing nitrogen and sulfur atoms with isothiazole isomer. Thiazoles are a basic scaffold found in many natural compounds as vitamin B1-thiamine, alkaloids, anabolic steroids, flavones [1].

The interest in the synthesis of compounds containing the thiazole moiety has been increasing steadily in view of their utility in the field of photosensitizers, rubber vulcanization [2], liquid crystals [3, 4], sensors [5], sunscreens [6], catalysts [7], dyes [8], pigments [1], and chromophores [9, 10]. Moreover, thiazoles occupy a prominent place in current medicinal chemistry due to their wide range of applications in the field of drug design and discovery [11]. They appear in the bacitracin, penicillin antibiotics [12], and various synthetic drugs as short-acting sulfa drug sulfathiazole [1]. Also, they are used as an antidepressant drug (pramipexole) [13], antiulcer agent (nizatidine) [14], anti-inflammatory drug (meloxicam) [15], HIV/AIDS drug (ritonavir) [16], and cancer treatment drug (tiazofurin) [17]. In fact, thiazole is a more common component of FDA-approved pharmaceuticals than related five-membered heterocycles such as isothiazole, thiophene, furan, isoxazole, and oxazole. On the other hand, the metal complexes of thiazole are widely used in photocatalysis [18]. 1,3-Thiazoles undergo different types of reactions to yield various biologically active fused heterocyclic moieties as thiazolopyrimidine, imidazothiazoles, thiazolopyridine, etc. [19–21].

## 2. Synthesis strategies of 1,3-thiazole derivatives

Thiazole ring system were easily synthesized by well-known methods of Hantzsch [22], Cook-Heilbron [23], and Gabriel [24]. A number of compounds may serve as nucleophilic reagent in this reaction, such as thioamides, thiourea, ammonium thiocarbamate or dithiocarbamate, and their derivatives. Hantzsch synthesized the simple thiazole nucleus in 1887 [25]. This synthesis approach involves cyclization and condensation of haloketones with thioamide, and it is considered the most widely popular process for the synthesis of thiazole moiety. In contrast, Gabriel synthesized thiazoles by treating α-acylaminoketones with stoichiometric amounts of P2S5 or Lawesson's reagent [26]. Also, Cook-Heilbron used versatile methods for the synthesis of substituted aminothiazoles involving the reaction of α-aminonitriles with dithioacids or esters, carbon disulfide, carbonyl sulfide, and isothiocyanates under mild conditions [27].

Lately, thiazole derivatives were synthesized in the presence of various catalysts [28–31] and with the use of a microwave irradiation technique [32].

### 2.1 Synthesis from α-halocarbonyl compounds (Hantzsch's synthesis) (type I)

#### *2.1.1 Reactions with thioamides*

Thioamides and various α-halocarbonyl compounds were reacted to give numerous thiazoles with alkyl, aryl, arylalkyl, or heteroaryl of several functional groups at position 2, 4, or 5 (**2.1.1**) [33, 34] (**Figure 1**).

#### *2.1.2 Reactions with N-substituted thiourea*

2-Monosubstituted or disubstituted aminothiazoles (**2.1.2**) were obtained by the reaction of halocarbonyl compounds with N-substituted thiourea compounds [35] (**Figure 2**).

#### *2.1.3 Reaction with esters of thiocarbamic acid*

The condensation of α-halocarbonyl compounds with thiocarbamates gave 2-hydroxythiazole derivatives (**2.1.3**) [36, 37] (**Figure 3**).

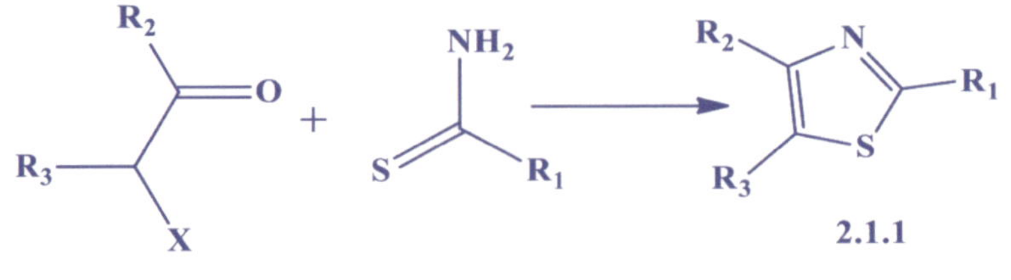


**2.1.1**

**Figure 1.**
*Synthesis of 2-, 4-, 5-trisubstituted thiazole.*

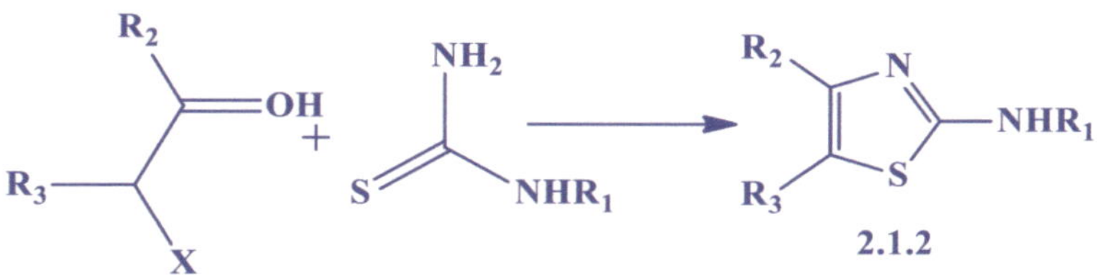


**Figure 2.**
*Synthesis of substituted aminothiazoles.*

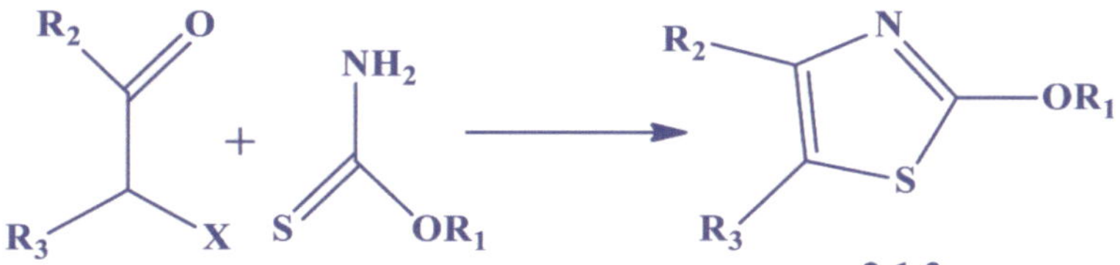


**Figure 3.**
*Synthesis of 2-hydroxythiazole derivatives.*

## 2.2 Synthesis from α-aminonitrile compounds (Cook-Heilbron's synthesis) (Type II)

This class of synthesis gives 5-aminothiazole with different substituted in position 2 by interacting aminonitrile with salts and esters of dithioacids carbon oxysulfide, carbon disulfide, and isothiocyanates significantly [38–40].

### *2.2.1 Reaction with carbon disulfide*

The condensation of carbon disulfide with α-aminonitriles gave 2-mercapto-5-amino thiazoles, which can be converted to 5-amino thiazoles substituted in position 2 (**2.2.1**) [41, 42] (**Figure 4**).

## 2.3 Reaction with esters and salts of dithioacids

The salts or the esters of both dithioformic and dithiophenacetic acids were reacted with α-aminonitriles to give 5-aminothiazoles (**2.3**) in good yields [43] (**Figure 5**).

## 2.4 Reaction with acylaminocarbonyl compounds and phosphorus pentasulfide and related condensation (Gabriel's synthesis) (Type III)

This reaction was originally designated by Gabriel in 1910. The reaction of phosphorus pentasulfide with acylaminoketone gave 2-phenyl-5-alkyl-thiazole in good yield (**2.4**) [44] (**Figure 6**).

## 2.5 Synthesis with eco-friendly methods

### *2.5.1 Using microwave-assisted synthesis (MAOS)*

The synthesis of thiazole derivatives involves vigorous reaction conditions and wastage of solvents and catalysts. To overcome these shortcomings, eco-friendly methods as microwave irradiation technique are commonly used for synthesis of

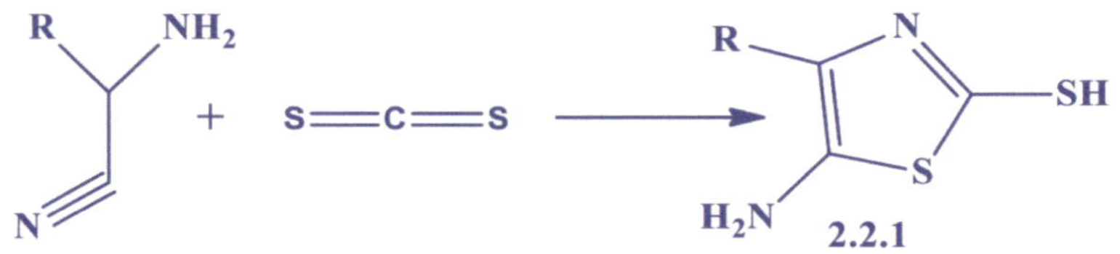


**Figure 4.**
*Synthesis of 5-aminothiazole derivatives.*

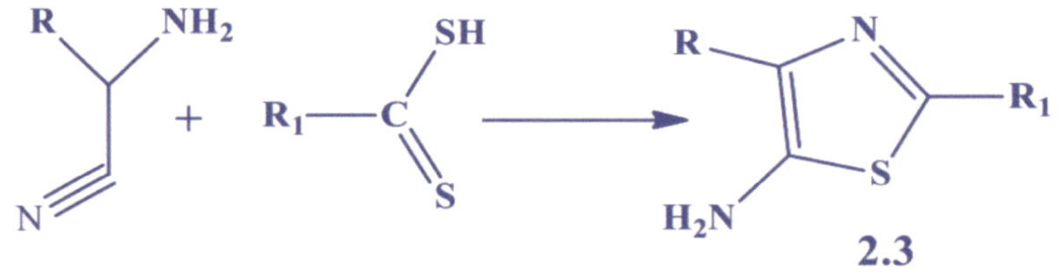


**Figure 5.**
*Synthesis of 5-aminothiazoles derivatives.*

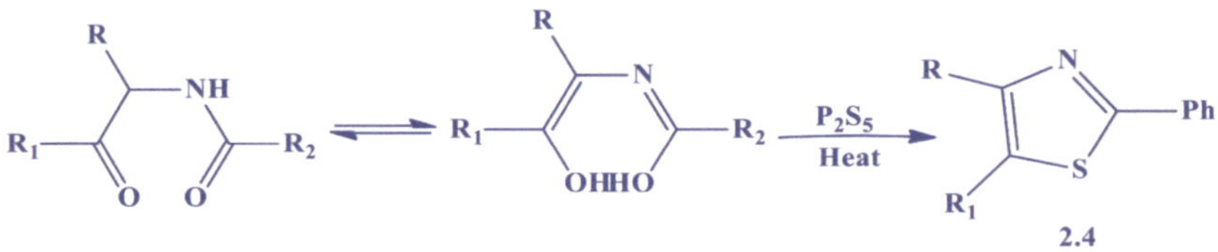


**Figure 6.**
*Synthesis of 2-phenyl-5-alkyl-thiazole derivatives.*

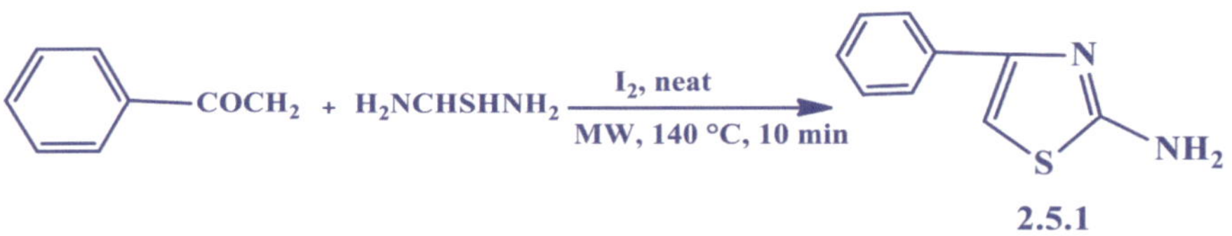


**Figure 7.**
*Synthesis of thiazoles under microwave irradiation.*

thiazole derivatives [45]. Rapid and elegant synthesis of a series of thiazoles (**2.5.1**) uses microwave heating under solvent-free conditions [32, 46, 47] (**Figure** 7).

### *2.5.2 One-pot multicomponent reaction in aqueous medium*

Water is economically viable, nontoxic, and the most friendly reaction medium available, making it an environmentally acceptable solvent for the design and development of green chemistry technique. A three-component reaction of phenyl acetylene, N-bromosuccinimide, and thiourea in aqueous medium gave substituted thiazole derivatives (**2.5.2**) in good yield [48] (**Figure 8**).

### *2.5.3 Using silica-supported tungstosilisic acid*

An efficient and green method has been developed for the synthesis of new substituted Hantzsch thiazole derivatives (**2.5.3**) by one-pot multicomponent procedure. 3-(Bromoacetyl)-4-hydroxy-6-methyl-2*H*-pyran-2-one was reacted with thiourea and substituted benzaldehydes in the presence of silica-supported tungstosilisic acid as a catalyst under conventional heating or under ultrasonic irradiation technique [46, 49] (**Figure 9**).

### 2.6 Miscellaneous methods

Hantzsch construction of thiazole derivatives (**2.6**) was established by the reaction of α-chloroglycinate esters with thioamides or thioureas. Targeted compounds are obtained from readily available and inexpensive building blocks through an environmentally benign process and without catalysts [50] (**Figure 10**).

The C – H substitution reaction of thiazole by the catalysis of the palladium/copper system is carried out in the presence of tetrabutylammonium fluoride under mild conditions. Various 2,5-diarylthiazole derivatives (**2.6.1**) were synthesized in good yields [51] (**Figure 11**).

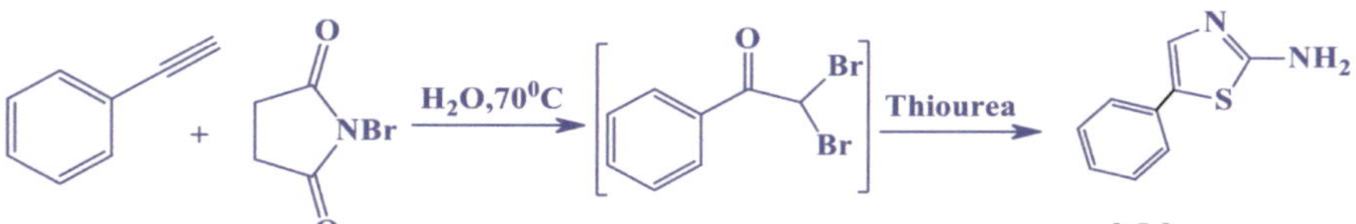


2.5.2

**Figure 8.**
*Synthesis of 2-aminothiazole in aqueous medium.*

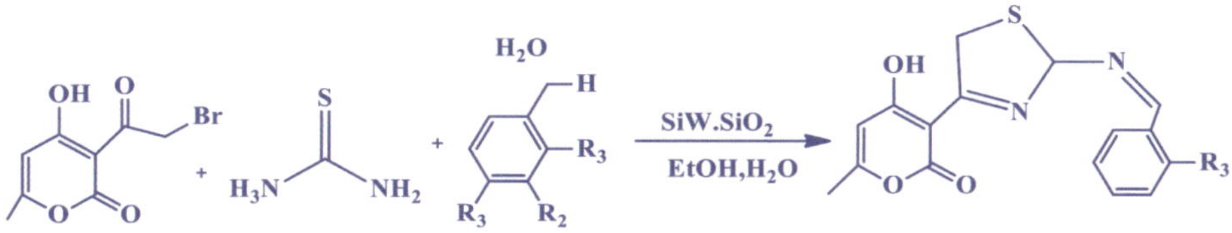


2.5.3

**Figure 9.**
*Synthesis of thiazole derivatives using silica.*

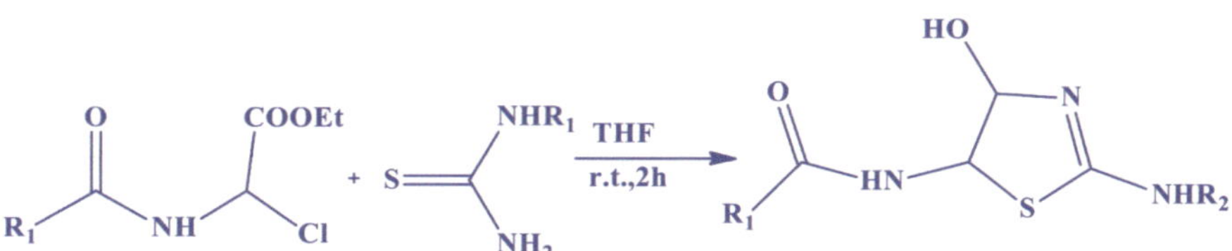


2.6

**Figure 10.**
*Synthesis of thiazole derivatives.*

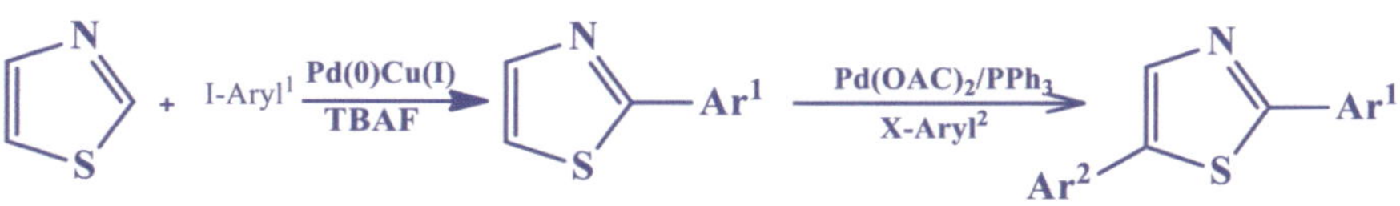


2.6.1

**Figure 11.**
*Synthesis of thiazole derivatives using palladium/copper.*

**Figure 12.**
*Synthesis of thiazole derivatives using oxidizing agents.*

### 2.7 Using oxidizing agents and thiourea

The mixtures of thiourea and acetophenone were treated with various oxidizing gents as sulfuryl chloride, chlorosulfonic acid, thionyl chloride, sulfur monochloride, sulfur trioxide, sulfuric acid, nitric acid, and sulfur. In each case a large amount of 2-amino-4-phenylthiazole (**2.7**) was obtained [52] (**Figure 12**).

## 3. Biological importance of thiazoles

Thiazole and its derivatives are among the most active classes of compounds that are known for their broad spectrum of activity, e.g., antibacterial [53], antifungal [54], antimalarial [55], antitubercular [56], antiviral [57], anti-inflammatory [58], antidiabetic [59], anthelmintic [60], anticonvulsant [61], antioxidant [62], anticancer [63], and cardiovascular activities [64], and known as new inhibitors of bacterial DNA gyrase B [65]. Some drugs that already are on the market including the recent entry dasatinib possess thiazoles nucleus [66].

### 3.1 Antitumor activity

Compounds containing thiazole have marked their presence in a number of clinically available anticancer drugs such as tiazofurin [67], dasatinib [68], dabrafenib [69], patellamide A [70], ixabepilone [71], and epothilone [72].

Ramla et al. synthesized a variety of 4-amino-3-methyl-5-(2-methyl-1*H*-benzo[*d*]imidazol-1-yl)thiazol-2(3*H*)-one (**3.1.1**) and evaluated them for antitumor activity [73] (**Figure 13**).

Popsavin et al. reported a set of 2-(2,3-anhydrofuranosyl) thiazole-4-carboxamide (2′,3′-anhydrotiazofurin) derivatives (**3.1.2**) and screened them for their antitumor activity [74] (**Figure 14**).

A series of 5-arylidene derivatives were synthesized and evaluated for their antitumor activity. Compound 2-{2-[3-(benzothiazol-2-ylamino)-4-oxo-2-thioxothiazolidin-5-ylidenemethyl]-4-chlorophenoxy}-N-(4-methoxyphenyl)-acetamide (**3.1.3**) was found to be the most active among the tested compounds [75] (**Figure 15**).

In another approach towards triple-negative breast cancer, Zhou et al. synthesized and optimized a series of hybrids of 2,4-diaminopyrimidine and thiazole derivatives (**3.1.4**). These compounds showed anti-proliferative properties against two breast cancer cell lines, MCF-7 and MDA-MB-231. Several of these compounds also exhibited potent activities against tumor cell colony [76] (**Figure 16**).

A series of 2-(4-benzoyl-phenoxy)-N-(4-phenyl-thiazol-2-yl)-acetamides were synthesized by Prashanth et al. The authors suggest that the effect of compound (**3.1.5**) could be due to methyl, fluoro, and methoxy groups which are attached to phenoxy, benzoyl, and the phenyl ring of thiazole, respectively [77] (**Figure 17**).

Dae-Kee K et al. produced a set of 5-(pyridin-2-yl)thiazoles enclosing a *p*- and/or *m*-carboxamide or carbonitrile-substituted phenylmethylamino moiety at position 2 of the thiazole ring (**3.1.6**). This series is evaluated for its ALK5 inhibitory activity [78, 79] (**Figure 18**).

A series of 2,4-disubstituted thiazole compounds containing *N*-n-butyl or *N*-cyclohexyl thioureido synthon at position 2 and N-substituted thiosemicarbazone moiety (**3.1.7**) at position 4 were synthesized by HI El-Subbagh et al. and verified for their antitumor activity. All of the established derivatives revealed antineoplastic activity [80] (**Figure 19**).

Santos et al. synthesized 6,7-bis(hydroxymethyl)-1H,3H-pyrrolo[1,2-c]thiazole (**3.1.8**) which showed activity for the triple-negative breast cancer, the most challenging tumor in clinical practice [81] (**Figure 20**).

El-Borai et al. synthesized a series of 2 6-substituted-3-(pyridin-3-yl) imidazo[2,1-b]thiazole (**3.1.9**) which are tested for anticancer activity against human cancer cell lines HEPG2 (liver cancer) and MCF7 using sulforhodamine B

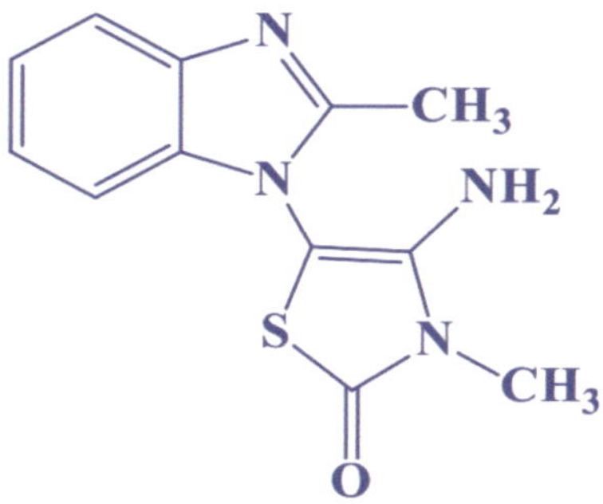


**Figure 13.**
*Structure of compound 3.1.1.*

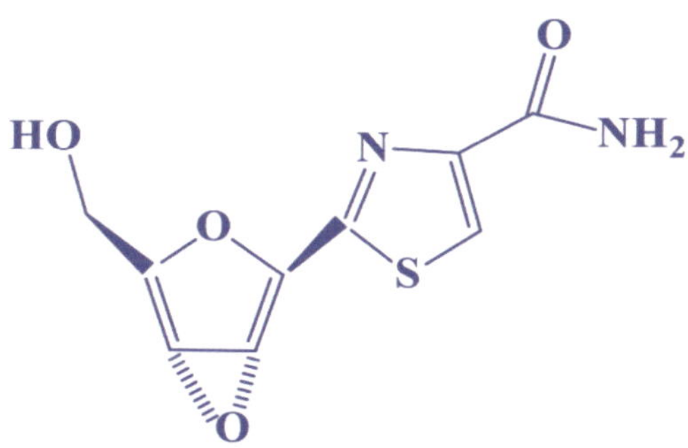


**Figure 14.**
*Structure of compound 3.1.2.*

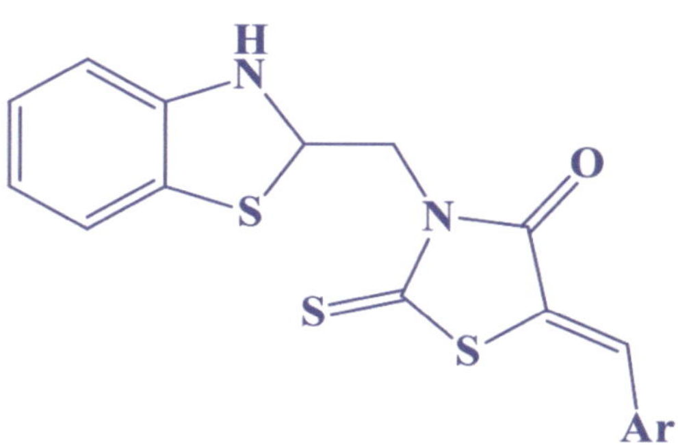


**Figure 15.**
*Structure of compound 3.1.3.*

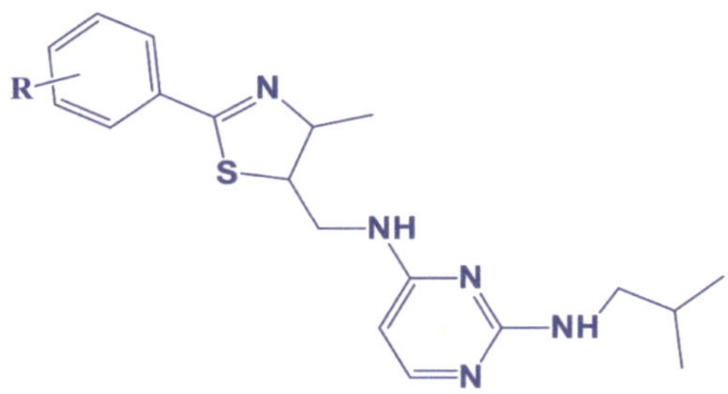


**Figure 16.**
*Structure of compound 3.1.4.*

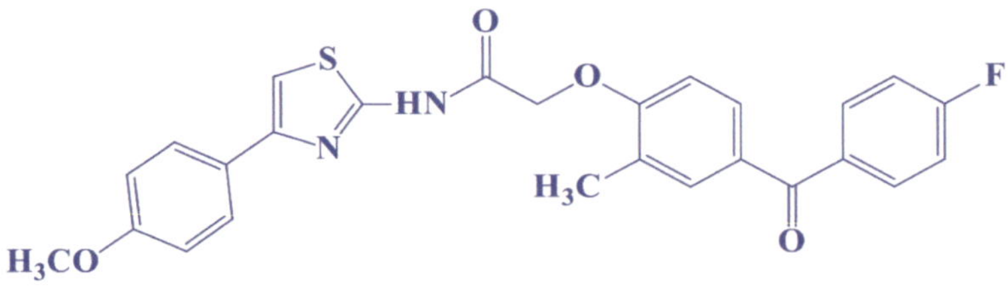


**Figure 17.**
*Structure of compound 3.1.5.*

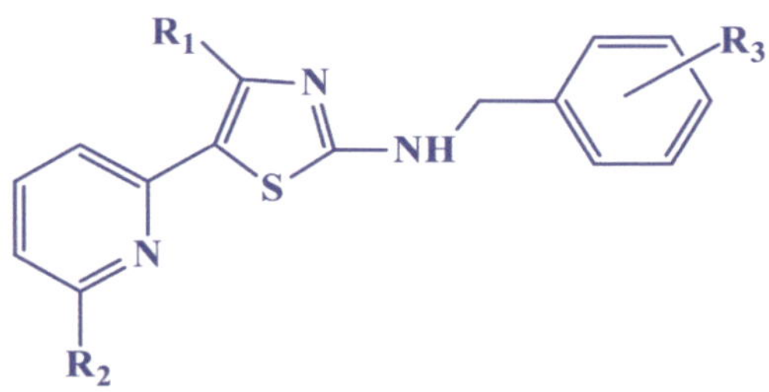


**Figure 18.**
*Structure of compound 3.1.6.*

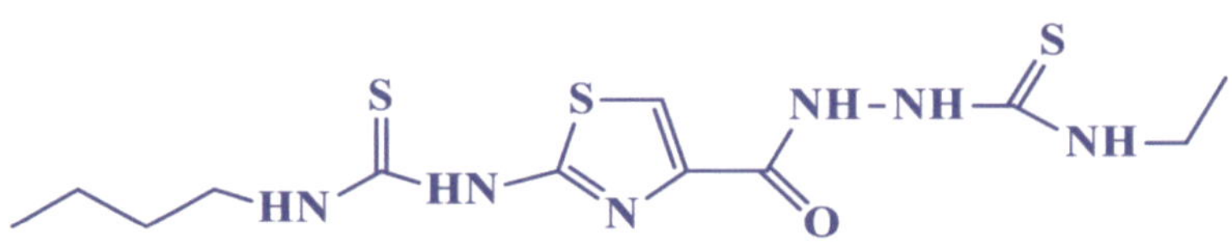


**Figure 19.**
*Structure of compound 3.1.7.*

(SRB) assay. All the synthesized compounds displayed more anticancer activity towards the selected cell line cancer, suggesting that it might be a potential alternative agent for human hepatic cancer therapy [82] (**Figure 21**).

### 3.2 Antimicrobial activity

Fungal and bacterial resistance to antimicrobial drugs is increasing rapidly due to nonselective antimicrobial activities and a limited number of drugs. To overcome this situation, several molecules containing thiazole are synthesized to treat bacterial and fungal infections [83, 84].

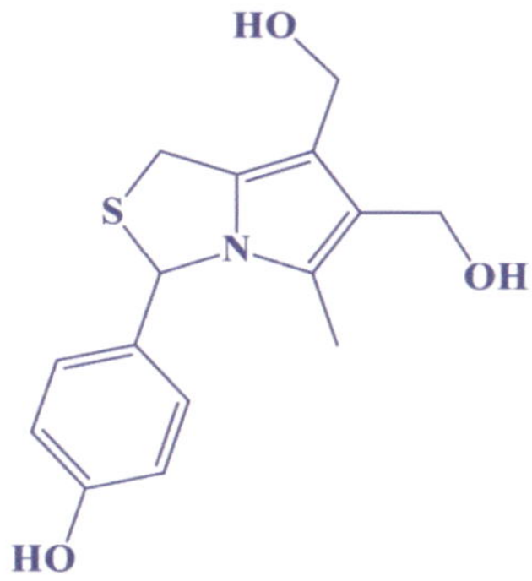


**Figure 20.**
*Structure of compound 3.1.8.*

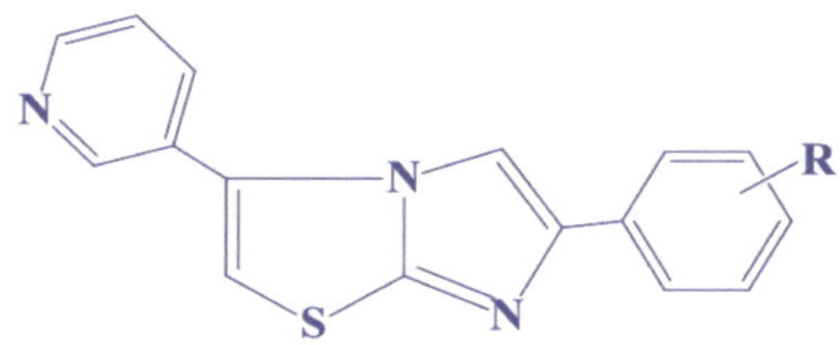


**Figure 21.**
*Structure of compound 3.1.9.*

El-Borai et al. work on an ongoing program in the field of synthesis and evaluated antimicrobial activity of medicinally important new compounds, taking the fused thiazole compounds as thiazolopyrimidines (**3.2.1**), imidazolothiazoles (**3.2.2**), and their derivatives as new examples in this domain [82] (**Figure 22**).

Vicini et al. synthesized a new set of 2-thiazolylimino-5-arylidene-4-thiazolidinones which were assayed in vitro for their antimicrobial activity against Gram-positive and Gram-negative bacteria and yeast. Compound (**3.2.3**) exhibited activity against Gram-positive bacteria [85] (**Figure 23**).

A series of thiazolyl thiazolidine-2,4-dione derivatives were synthesized by Dundar et al. These compounds were screened for their antibacterial and antifungal activities against methicillin-resistant *S. aureus*, *E. coli*, and *C. albicans*. All the compounds particularly (**3.2.4**) were found to be moderately potent against screened microorganisms [86] (**Figure 24**).

Abdel-Wahab et al. synthesized 3-(benzofuran-2-yl)-4,5-dihydro-5-aryl-1-[4-(aryl)-1,3-thiazol-2-yl]-1*H*-pyrazoles (**3.2.5**). The synthesized compounds were screened for their antibacterial and antifungal activities and showed a significant activity against *E. coli* higher than that of the control drug, whereas antifungal activity against *Aspergillus niger* was also exhibited and equal to that of the reference drug [87] (**Figure 25**).

Bera et al. Synthesized pyridinyl thiazole ligand having hydrazone moiety and its cobalt complex. Both ligand and its complex were tested for antibacterial properties towards Gram-positive and Gram-negative bacteria. The results revealed that the ligand (**3.2.6**) exhibited excellent antibacterial activity. The presence of pyridinium ion in the ligand showed increased solubility of the ligand which enhances the cell penetrating ability and cell binding activity of the ligand. Hydrolysis of ligand decreases the pH of the medium which facilitates easy penetration of ligand into the cell [88] (**Figure 26**).

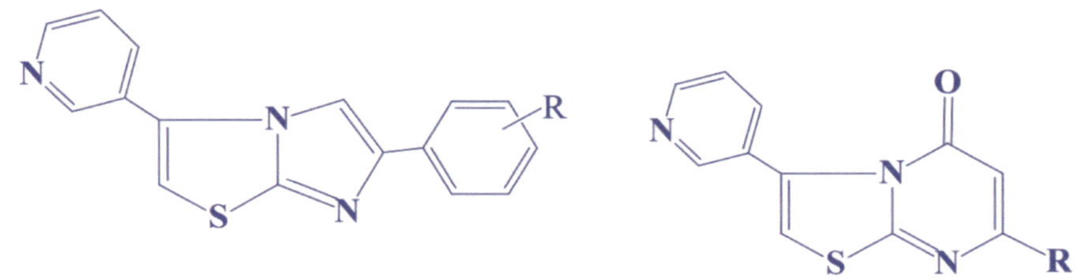


**Figure 22.**
*Structure of compounds 3.2.1 and 3.2.2.*

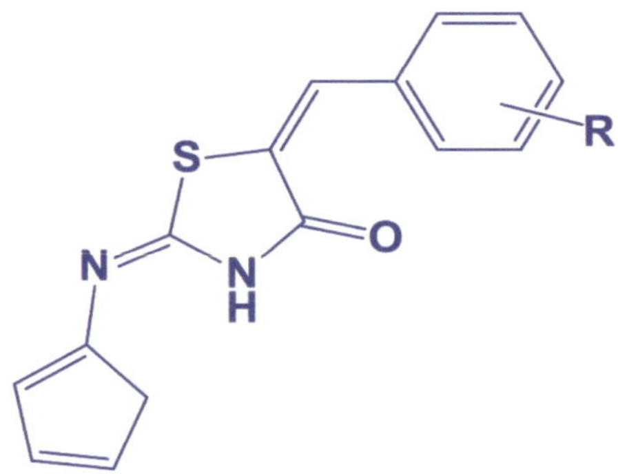


**Figure 23.**
*Structure of compound 3.2.3.*

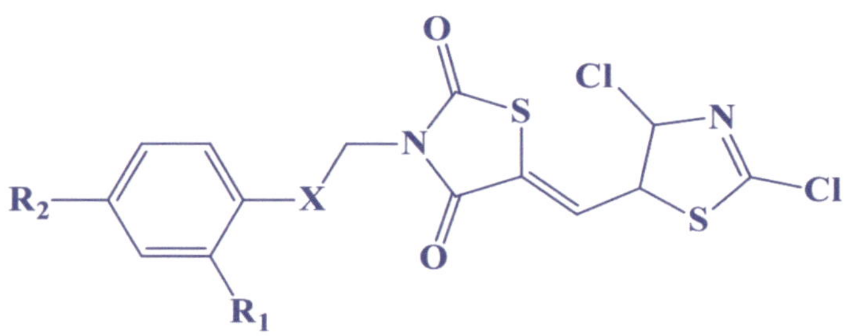


**Figure 24.**
*Structure of compound 3.2.4.*

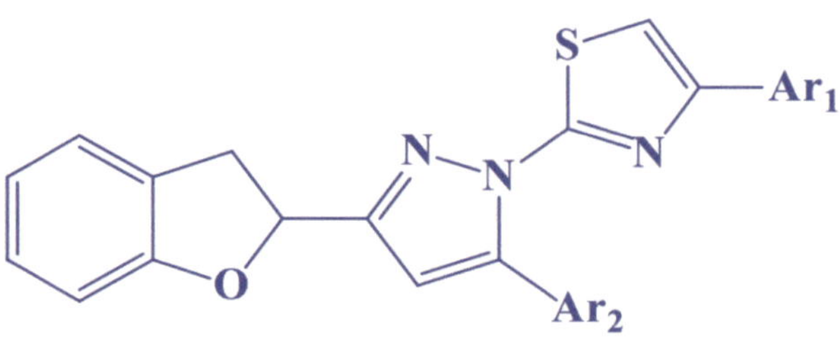


**Figure 25.**
*Structure of compound 3.2.5.*

### 3.3 Antifungal activity

Narayana et al. synthesized a series of 5-(2-substituted–1,3-thiazol-5-yl)-2-alkoxybenzamides and 5-(2-*N*-(substituted aryl)-1,3-thiazol-5-yl)-2-alkoxy benzamides. The synthesized compounds were screened for their antifungal activity. The derivatives of compound (**3.3.1**) exhibited significant activity [89] (**Figure 27**).

Chimenti et al. reported the synthesis of a novel series of 2-thiazolylhydrazone derivatives and the influence of the substituents on the thiazole ring and on antifungal activity. Some of the tested compounds were found to possess significant antifungal activity when compared to clotrimazole, in particular compound (**3.3.2**) which exhibited higher potency against most of the *Candida* [90] (**Figure 28**).

### 3.4 Antioxidant activity

Antioxidants are of great interest due to their participation in important biological and industrial processes. They are generated in the human body and may cause damage to lipids, proteins, and DNA and thus may lead to various diseases such as cancer, atherosclerosis, diabetes, cirrhosis, and Alzheimer's and inflammatory diseases [91]. Thiazole and derivatives are the core structure in a variety of pharmaceuticals with a wide range of biological activity [92–94].

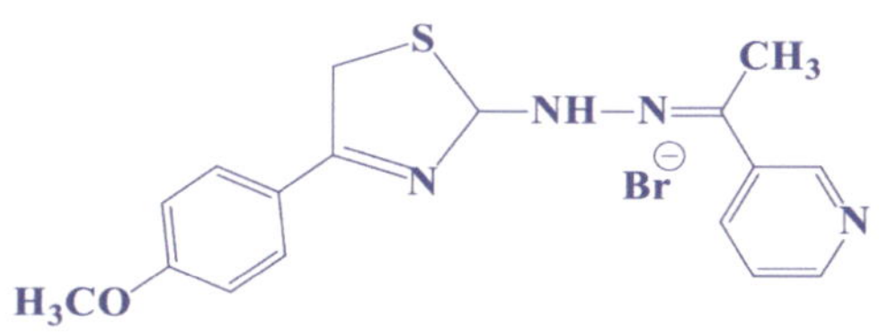


**Figure 26.**
*Structure of compound 3.2.6.*

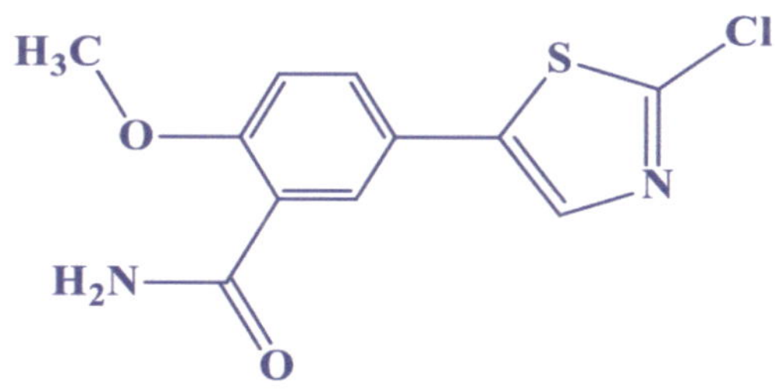


**Figure 27.**
*Structure of compound 3.3.1.*

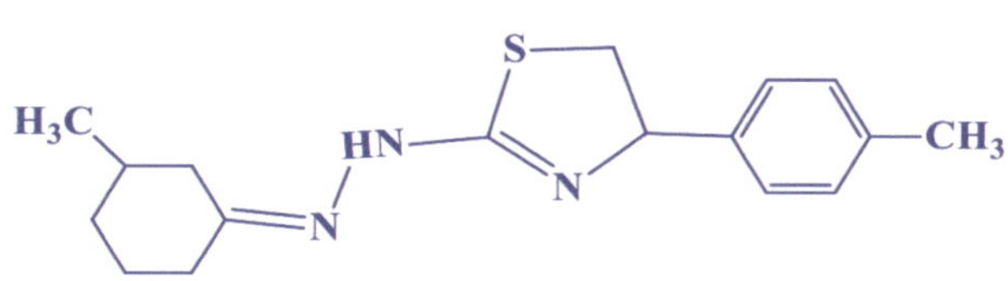


**Figure 28.**
*Structure of compound 3.3.2.*

The antioxidant potential compounds (**3.4.1**) was evaluated by spectrophotometric method, using DPPH radical or Fe $(TPTZ)^{3+}$ complex, and EPR spectroscopy and revealed that the synthesized compounds were showing potent antioxidant activity [95] (**Figure 29**).

Bozdag-Dundar et al. synthesized a series of 2, 4-dichlorothiazolyl thiazolidine-2,4-dione and 4-chloro-2-benzylsulfanylthiazolyl-thiazolidine-2,4-dione derivatives, and they were tested for their antioxidant properties. Compound (**3.4.2**) showed the best superoxide anion scavenging activity [96] (**Figure 30**).

Gouda et al. synthesized 2-amino thiazole derivatives and evaluated their antioxidant activity. They reported that the three compounds (**3.4.3**) showed potent antioxidant activity after postulating the structure–activity relationship (SAR) [97] (**Figure 31**).

A series of N2-[2-chloro-4(3,4,5-trimethoxy phenyl) azetidin-1-yl)]-N4-(substituted aryl)-1,3-thioazol-2,4-diamine (**3.4.4**) were synthesized and screened for their in vitro antioxidant properties. The IC50 values revealed that some of the synthesized compounds were showing potent antioxidant activity [98] (**Figure 32**).

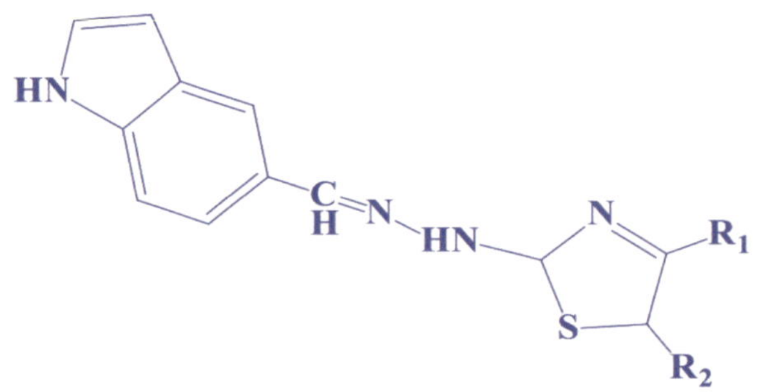


**Figure 29.**
*Structure of compound 3.4.1.*

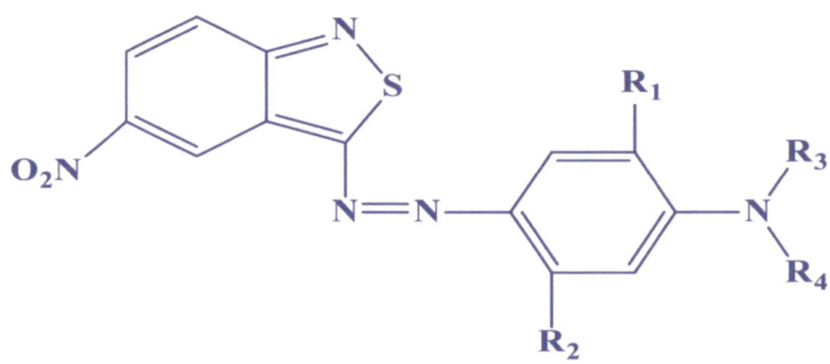


**Figure 30.**
*Structure of compound 3.4.2.*

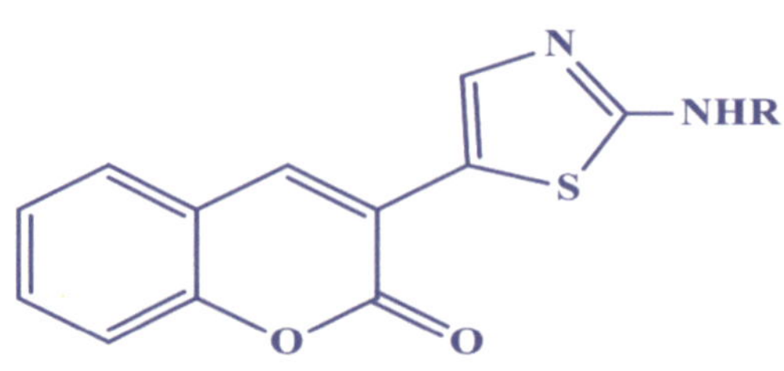


**Figure 31.**
*Structure of compound 3.4.3.*

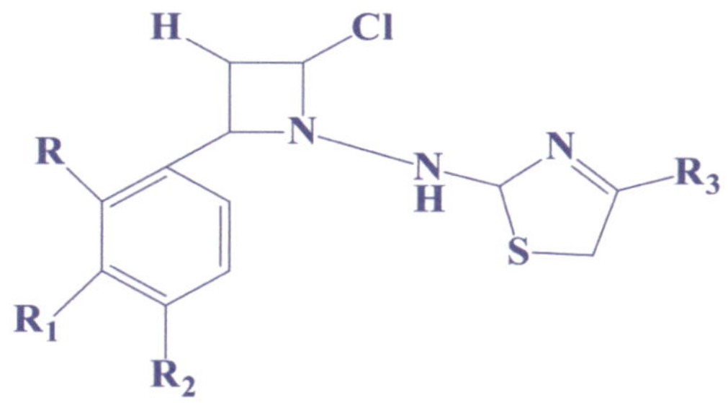

**Figure 32.**
*Structure of compound 3.4.4.*

## 4. Conclusion

Thiazole moieties have occupied a pivotal position in the modern organic and medicinal chemistry due to its broad-spectrum pharmacological and medicinal activities such as antimicrobial, anticancer, and antioxidant. The presence of thiazole ring in many drugs such as penicillin, pramipexole, tiazofurin, meloxicam, and nizatidine motivates the chemists to design new thiazole scaffolds. Thiazole nucleus exhibited an important role in finding new leads and drugs for various diseases. This chapter has illustrated the commonly used approaches to synthesize subsisted thiazole derivatives, described their key electronic properties, and highlighted their most important chemical reactivity. A particular focus has been on the use of thiazole in dyes and their metal complexes and miscellaneous applications of thiazole dyes. Also we have focused our attention on the biological application of thiazole derivatives.

## List of abbreviations

| | |
|---|---|
| FDA | Food and Drug Administration (USA) |
| SAR | structure–activity relationships |
| MAOS | microwave-assisted synthesis |
| HTIB | [hydroxy-(tosyloxy)-iodo] benzene |
| TBAF | tetrabutylammonium fluoride |

## Author details

Seham A. Ibrahim* and Hala F. Rizk
Department of Chemistry, Faculty of Science, Tanta University, Tanta, Egypt

*Address all correspondence to: sehamabdelatif@yahoo.com

## References

[1] Borisenko VE, Koll A, Kolmakov EE, Rjasnyi AG. Hydrogen bonds of 2-aminothiazoles in intermolecular complexes (1,1 and 1,2) with proton acceptors in solutions. Journal of Molecular Structure. 2006;**783**(1):101-115

[2] Milne GWA. Handbook of Antineoplastic Agents. London, UK: Gower/Ashgate; 2000

[3] Kiryanov AA, Sampson P, Seed AJ. Synthesis of 2-alkoxysubstituted thiophenes, 1,3-thiazoles, and related S-heterocycles *via* Lawesson's reagent-mediated cyclization under microwave irradiation: Applications for liquid crystal synthesis. The Journal of Organic Chemistry. 2001;**66**:7925

[4] Mori A, Sekiguchi A, Masui K, Shimada T, Horie M, Osakada K, et al. Facile synthesis of 2,5-diarylthiazoles via palladium-catalyzed tandem C–H substitutions. Design of tunable light emission and liquid crystalline characteristics. Journal of the American Chemical Society. 2003;**125**:1700

[5] Kim BY, Kim HS, Helal A. A fluorescent chemosensor for sequential recognition of gallium and hydrogen sulfate ions based on a new phenylthiazole derivative. Sens. Actuators B-Chem. 2015;**206**:430

[6] Bach T, Heuser S. Synthesis of 2-(*o*-hydroxyaryl)-4-arylthiazoles by regioselective Pd(0)-catalyzed cross-coupling. Tetrahedron Letters. 2000;**41**:1707

[7] Dondoni A. Heterocycles in organic synthesis: Thiazoles and triazoles as exemplar cases of synthetic auxiliaries. Organic & Biomolecular Chemistry. 2010;**8**:3366

[8] Metwally MA, Abdel-Latif E, Amer FA, Kaupp G. Synthesis of new 5-thiazolyl azo-disperse dyes for dyeing polyester fabrics. Dyes and Pigments. 2004;**609**(3):249-264

[9] Guo XG, Quinn J, Chen ZH, Usta H, Zheng Y, Xia Y, et al. Dialkoxybithiazole: A New building block for head-to-head polymer semiconductors. Journal of the American Chemical Society. 2013;**135**:1986-1996

[10] Maj J, Rog Z, Skuza G, Kolodziejezyk K. Antidepressant effects of pramipexole, a novel dopamine receptor agonist. Journal of Neural Transmission. 1997;**104**(4-5):525-533

[11] VishnuJi R, Arun S, Mahendra N, Ramendra P. The Chemistry of Heterocycles: Nomenclature and Chemistry of Three-to-Five Membered Heterocycles2019. pp. 149-478

[12] Maulard T, Lagorce JF, Thomos JC, Raby C. Biological evaluation of compounds with –NCS group or derived from thiazole and imidazole. activity on prostaglandin synthetase complex. The Journal of Pharmacy and Pharmacology. 1993;**54**(8):731-735

[13] Markus B, Salome VG, Christoph B, Werner JP. Molecular aspects of drug recognition by specific T cells. Curr. Drug. Tar. 2004:1-11

[14] Knadler MP, Bergstrom RF, Callaghan JT, Rubin A. Nizatidine, an H2-blocker. Its metabolism and disposition in man. Drug Metabolism and Disposition. 1986;**14**(2):175-182

[15] Rehman MZ, Anwar CJ, Ahmad S. An Efficient synthesis of 2-alkyl-4-hydroxy-2H-1,2-benzothiazine-3-carboxamide-1,1-dioxides. Bulletin of the Korean Chemical Society. 2005;**26**(11):1771-1775

[16] De Souza MVN, De Almeida MV. Drugs anti-HIV: Past, present and future perspectives. Quimica Nova. 2003:366-372

[17] Popsavin M, Torović L, Svircev M. and antiproliferative activity of two new tiazofurin analogues with 2′-amido functionalities. Bioorganic & Medicinal Chemistry Letters. 2006;**16**(10):2773-2776

[18] Huo J, Zeng H. A novel triphenylamine functionalized bithiazole– metal complex with C60 for photocatalytic hydrogen production under visible light irradiation. Journal of Materials Chemistry. 2015;**3**:6258

[19] Koppireddi S, Chilaka DRK, Avula S, Komsani JR, Kotamraju S, Yadla R. Synthesis and anticancer evaluation of 3-aryl-6-phenylimidazo [2,1-b]thiazoles. Bioorganic & Medicinal Chemistry Letters. 2014;**24**:5428-5431

[20] Chaniyara R, Tala S, Chen C-W, Lee PC, Kakadiya R, Dong H, et al. Synthesis and antitumor evaluation of novel benzo[d]pyrrolo[2,1-b]thiazole derivatives. European Journal of Medicinal Chemistry. 2012;**53**:28-40

[21] Lee T, Lee D, Lee IY, Gong YD. Solid-phase synthesis of thiazolo[4,5-*b*] pyridine derivatives using friedländer reaction. Journal of Combinatorial Chemistry. 2010;**12**:95-99

[22] Serpil E, Emine TC, Onder İ, Ersin İ, Zehra K, Mısıra E, et al. Derivatives of pyridine and thiazole hybrid: Synthesis, DFT, biological evaluation *via* antimicrobial and DNA cleavage activity. Bioorganic & Medicinal Chemistry. 2020;**95**:103476

[23] Sammes PG. Comprehensive Organic Chemistry Heterocyclic Compounds, Part 20-1. Vol. 4. Oxford: Pegammon Press; 1979. p. 976

[24] Elederfield RC. Heterocyclic Compounds (Five Membered Heterocycles Containing Two Hetero-Atoms). New York: John Wiley and Sons Inc; 1961

[25] ZoltewiczJA,DeadyLW.Quaternization of heteroaromatic compounds: Quantitative Aspects. Advances in Heterocyclic Chemistry. 1978:71-121

[26] Gabriel S. Synthese von oxazolen und thiazolen II. European Journal of Inorganic Chemistry. 1910;**43**:1283-1287

[27] Rajappa S, Nair MD, Advani BG, Sreenivasan R, Desai JAA. General synthesis of thiazoles. Part 3. Comparative evaluation of different functionalised thioureas as precursors. Journal of the Chemical Society, Perkin Transactions. I. 1979:1762-1764

[28] Das B, Reddy SV, Ramu R. A rapid and high-yielding synthesis of thiazoles and aminothiazoles using ammonium-12-molybdophosphate. Journal of Molecular Catalysis A: Chemical. 2006:235-237

[29] Narender M, Reddy SM, Sridhar R, Nageswar YVD, Rao RK. Aqueous phase synthesis of thiazoles and aminothiazoles in the presence of β-cyclodextrin. Tetrahedron Letters. 2005:5953-5955

[30] Karade H, Sathe M, Kaushik MP. An efficient method for the synthesis of 2-aminothiazoles using silica chloride as a heterogeneous catalyst. Catalysis Communications. 2007;**8**:741-746

[31] Egido EG, Wong SYF, Warrington BH. A Hantzsch synthesis of 2-aminothiazoles performed in a heated microreactor system. Lab on a Chip. 2002;**2**(1):31-33

[32] George WK, Arjun RM. Microwave promoted synthesis of functionalized 2-aminothiazoles. Tetrahedron Letters. 2006;**47**:5171-5172

[33] Willstatter R, Wirth T. Über Thioformamid. Berichte: IntechOpen; 1909. pp. 1908-1922. DOI: 10.1002/9780470187081

[34] Otto H. Process for the Production of Thiazoles. U. S. Patent No. 2160867. ExLi4EvA; 1939

[35] George YS, Subhi AA. Preparation and spectral characterization of substituted 2-aminothiazoles. Journal of Chemical & Engineering Data. 1973;**18**(1):99-102

[36] Ayman WE, Sherif MS, Hatem MG. The chemistry of α-haloketones and their utility in heterocyclic synthesis. Molecules. 2003;**8**(1):793-865

[37] Someshwar P. Significance of Thiazole-Based Heterocycles for Bioactive Systems. IntechOpen; 2016. DOI: 10.5772/62077

[38] Cook AH, Heilbron I, MacDonald SF, Mahadevan AP. Studies in the azole series. Part XII. Some thiazolopyrimidines. The Royal Society of Chemistry's Journals, Books and Databases. 1949:1064-1068. DOI: 10.1039/JR9490001064

[39] Cook AH, Heilbron I, Mahadevan AP. Studies in the azole series. Part XI. The interaction of α-amino-nitriles, hydrogen sulphide, and ketones. Journal of the Chemical Society. 1949;**225**:1061-1064

[40] Mara T, Gabriele L, Pamela P, Federico VR, Samuele L, Gianluca B, et al. Catalyst-Free Synthesis of polysubstituted 5-acylamino-1,3-thiazoles via hantzsch cyclization of α-chloroglycinates. Molecules. 2019;**24**:3846

[41] Cook AH, Heilbron IM, Levy AL. Studies in the azole series. Part II. The interaction of α-amino-nitriles and carbon disulphide. Journal of the Chemical Society. 1947:1598-1609

[42] Cook AH, Heilbron IM, Stern E. Studies in the azole series. Part X. Some 5-amino-2-mercapto-4-alkylthiazoles and 2: 4-dithio-5-alkylhydantions. Journal of the Chemical Society. 1948:2031-2033

[43] Hoggarth E. The rearrangement of 2-benzenesulphenamidothiazoles. Part II. Thiazole compounds substituted in position 4. Journal of the Chemical Society. 1947:114-118

[44] Jacques VM. General synthetic methods for thiazole and thiazolium salts. The chemistry of heterocyclic compounds. In: Thiazole and Its Derivatives, Chapter II. Vol. 34. 1979

[45] Sukanta K, Kimberly M, Edward RB. Microwave-assisted Hantzsch thiazole synthesis of N-phenyl-4-(6-phenylimidazo[2,1-b] thiazol-5-yl)thiazol-2-amines from the reaction of 2-chloro-1-(6-phenylimidazo[2,1-b]thiazol-5-yl) ethanones and thioureas. Tetrahedron Letters. 2012;**53**:4921-4924

[46] Nayak S, Gaonkar SL. A Review on recent synthetic strategies and pharmacological importance of 1,3-thiazole derivatives. Mini-Reviews in Medicinal Chemistry. 2019;**19**(3):215-238

[47] Cáceres-Castillo D, Carballo RM, Tzec-Interián JA, Mena-Rejón GJ. Solvent-free synthesis of 2-amino-4-arylthiazoles under microwave irradiation. Tetrahedron Letters. 2012;**53**(30):3934-3936

[48] Madhav B, Murthy SN, Anil Kumar BSP, Ramesh K, Nageswar YVD. A tandem one-pot aqueous phase synthesis of thiazoles/selenazoles. Tetrahedron let. 2012;**53**:3835-3838

[49] Bouherrou H, Saidoun A, Abderrahmani A, Abdellaziz L, Rachedi Y, Dumas F, et al. Synthesis and biological evaluation of new substituted hantzsch thiazole derivatives from environmentally benign one-pot synthesis Using silica supported tungstosilisic acid as reusable catalyst. Molecules. 2017;**22**(5):E757

[50] Tomassetti M, Lupidi G, Piermattei P, Rossi FV, Lillini S, Bianchini G, et al.

Catalyst-free synthesis of polysubstituted 5-acylamino-1, 3-thiazoles via hantzsch cyclization of α-chloroglycinates. Molecules. 2019;**24**(21):3846

[51] Mori A, Sekiguchi A, Masui K, Shimada T, Horie M, Osakada K, et al. Facile synthesis of 2,5-diarylthiazoles via palladium-catalyzed tandem C-H substitutions. Design of tunable light emission and liquid crystalline characteristics. Journal of the American Chemical Society. 2003;**125**(7):1700-1701

[52] Dodson RM, King LC. The reaction of acetophenone with thiourea and oxidizing agents. Journal of the American Chemical Society. 1946;**68**(5):871-871

[53] Waad DA, Mahmoud EH, Shahenda MM, Abdulmalik A, Hussein S, El-Sayed EH. Antibacterial, antibiofilm and molecular modeling study of some antitumor thiazole based chalcones as a new class of DHFR inhibitors. Microbial Pathogenesis. 2019;**136**:103674

[54] Khabnadideh S, Rezaei Z, Pakshir K, Zomorodian K, Ghafari N. Synthesis and antifungal activity of benzimidazole, benzotriazole and aminothiazole derivatives. Research in Pharmaceutical Sciences. 2012;**7**(2):65-72

[55] Bueno JM, Carda M, Crespo B, Cunat AC, de Cozar C, Leon ML, et al. Design, synthesis and antimalarial evaluation of novel thiazole derivatives. Bioorganic & Medicinal Chemistry Letters. 2016;**26**(16):3938-3944

[56] Andreani A, Granaiola M, Leoni A, Locatelli A, Morigi R, Rambaldi M. Synthesis and antitubercular activity of imidazo[2,1-b]thiazoles. European Journal of Medicinal Chemistry. 2001;**36**(9):743-746

[57] Dawood KW, Eldebss TM, El-Zahabi HS, Yousef MH. Synthesis and antiviral activity of some new bis-1,3-thiazole derivatives. European Journal of Medicinal Chemistry. 2015;**18**(102):266-276

[58] Sharma RN, Xavier FP, Vasu KK, Chaturvedi SC, Pancholi SS. Synthesis of 4-benzyl-1, 3-thiazole derivatives as potential anti-inflammatory agents: An analogue-based drug design approach. Journal of Enzyme Inhibition and Medicinal Chemistry. 2009;**24**(3):890-897

[59] Bozdag-Dundar O, Ceylan-Unlusoy M, Verspohl EJ, Ertan R. Synthesis and antidiabetic activity of novel 2,4-thiazolidinedione derivatives containing a thiazole ring. Arzneimittel-Forschung/Drug Research. 2006;**56**(9):621-625

[60] Weikert RJ, Bingham S Jr, Emanuel MA, Fraser-Smith EB, Loughhead DG, Nelson PH, et al. Synthesis and anthelmintic activity of 3′-benzoylurea derivatives of 6-phenyl-2,3,5,6-tetrahydroimidazo[2,1-b] thiazole. Journal of Medicinal Chemistry. 1991;**34**(5):1630-1633

[61] Ucar H, Van Derpoorten K, Cacciaguerra S, Spampinato S, Stables JP, Depovere P, et al. Synthesis and anticonvulsant activity of 2(3H)-benzoxazolone and 2(3H)-benzothiazolone derivatives. Journal of Medicinal Chemistry. 1998;**41**(7):1138-1145

[62] Kurt BZ, Gazioglu I, Sonmez F, Kucukislamoglu M. Synthesis, antioxidant and anticholinesterase activities of novel coumarylthiazole derivatives. Bioorganic Chemistry. 2015;**59**:80-90

[63] Thoraya AF, Ghada SM, Zeinab AM, Marwa FH. Discovery of thiazole-based-chalcones and 4-hetarylthiazoles as potent anticancer agents: Synthesis, docking study and anticancer activity. Bioorganic Chemistry. 2020;**98**:103761

[64] Omar AM, Eshba NH. Synthesis and biological evaluation of new

2,3-dihydrothiazole derivatives for antimicrobial, antihypertensive, and anticonvulsant activities. Journal of Pharmaceutical Sciences. 1984;**73**(8):1166-1168

[65] Xu R, Tian Y, Huang S, Yu J, Deng Y, Zhan M, et al. Synthesis and evaluation of novel thiazole-based derivatives as selective inhibitors of DNA-binding domain of the androgen receptor. Chemical Biology & Drug Design. 2018;**91**(1):172-180

[66] Das D, Sikdar P, Bairagi M. Recent developments of 2-aminothiazoles in medicinal chemistry. European Journal of Medicinal Chemistry. 2015;**109**:89-98. DOI: 10.1016/j.ejmech.2015.12.022

[67] Franchetti P, Cappellacci L, Grifantini M, Barzi A, Nocentini G, Yang H, et al. Furanfurin and thiophenfurin: Two novel tiazofurin analogues. Synthesis, structure, antitumour activity, and interactions with inosine monophosphate dehydrogenase. Journal of Medicinal Chemistry. 1995;**38**(19):3829-3837

[68] Li X, He Y, Ruiz CH, Koenig M, Cameron MD. Characterization of dasatinib and its structural analogs as CYP3A4 mechanism-based inactivators and the proposed bioactivation pathways. Drug Metabolism and Disposition. 2009;**37**(6):1242-1250

[69] Hu-Lieskovan S, Mok S, Homet Moreno B, Tsoi J, Robert L, Goedert L, et al. An Improved antitumour activity of immunotherapy with B-RAF and MEK inhibitors in BRAF(V600E) melanoma. Science Translational Medicine. 2015;**18**(7):279, 279-41

[70] Rashid MA, Gustafson KR, Cardellina JH, Boyd MR, Patellamide F. A new cytotoxic cyclic peptide from the colonial ascidian *Lissoclinum patella*. Journal of Natural Products. 1995;**58**(4):594-597

[71] Yao Y, Chen S, Zhou X, Xie L, Chen A. 5-FU and ixabepilone modify the microRNA expression profiles in MDA-MB-453 triple-negative breast cancer cells. Oncology Letters. 2014;**7**:541-547

[72] Altmann KH. Epothilone B and its analogs—A new family of anticancer agents. Mini Reviews in Medicinal Chemistry. 2003;**3**(2):149-158

[73] Ramla MM, Omar MA, El-Khamry AMM, El-Diwan HI. Synthesis and antitumor activity of 1-substituted-2-methyl-5-nitrobenzimidazoles. Bioorganic & Medicinal Chemistry. 2006;**14**:7324

[74] Popsavin M, Spaic S, Svircev M, Kojic V, Bogdanovic G, Popsavin V. Synthesis and antitumor activity of new tiazofurin analogues bearing a 2,3-anhydro functionality in the furanose ring. Bioorganic & Medicinal Chemistry Letters. 2007;**17**:4123

[75] Helal CJ, Sanner MA, Cooper CB, Gant T, Adam M, Lucas JC, et al. Discovery and SAR of 2-aminothiazole inhibitors of cyclin-dependent kinase 5/p25 as a potential treatment for Alzheimer's disease. Bioorganic & Medicinal Chemistry. 2004;**14**:5521

[76] Mohammadi-Farani A, Foroumadi A, Rezvani Kashani M, Aliabadi A. *N*-Phenyl-2-*p*-tolylthiazole-4-carboxamide derivatives: Synthesis and cytotoxicity evaluation as anticancer agents. Iranian Journal of Basic Medical Sciences. 2014;**17**:502-508

[77] Zhou W, Huang A, Zhang Y, Lin Q, Guo W, You Z, et al. Design and optimization of hybrid of 2,4-diaminopyrimidine and arylthiazole scaffold as anticancer cell proliferation and migration agents. European Journal of Medicinal Chemistry. 2015;**96**:269-280

[78] Gross S, Lengauer C, Hoeflich KP, Gross S, Rahal R, Stransky N, et al.

Targeting cancer with kinase inhibitors. Journal of Clinical Investigation. 2015;**125**:1780-1789

[79] Kim D, Choi JH, An J, Soon H. Synthesis and biological evaluation of 5-(pyridin-2-yl) thiazoles as transforming growth factor-b type1 receptor kinase inhibitors. Bioorganic & Medicinal Chemistry Letters. 2008;**18**:2122-2212

[80] Kojic V, Svirc M, Bogdanovic G. Synthesis and in vitro antitumour screening of 2-(B-D-xylofuranosyl) thiazole-4-carboxamide and two novel tiazofurin analogues with substituted tetrahydrofurodioxol moiety as a sugar mimic. Bioorganic & Medicinal Chemistry Letters. 2012:226700-226704

[81] Virginia S, Alessandro A, Stella C, Anna C, Alessandra M, Paola B, et al. Synthesis and antitumor activity of new thiazole nortopsentin analogs. Marine Drugs. 2016;**14**(12):226

[82] dos Santos TAR, da Silva AC, Silva EB, de Moraes Gomes PAT, Espíndola JWP, de Oliveira Cadoso MV, et al. Antitumor and immunomodulatory activities of thiosemicarbazones and 1,3-thiazoles in Jurkat and HT-29 cells. Biomed, Pharma. 2016;**82**:555-556

[83] Lin R, Johnson SG, Connolly PJ, Wetter SK, Binnun E, Hughes TV, et al. Analogues as antitumor epidermal growth factor receptor (EGFR) tyrosine kinase inhibitors. Bioorganic & Medicinal Chemistry Letters. 2009;**19**:2333-2337

[84] Liu W, Zhang X, Zhao J, Li Cui J, Mao Z. Inhibition of cervical cancer cell metastasis by benzothiazole through up-regulation of E-cadherin expression. Microbial Pathogenesis. 2017;**111**:182-186

[85] Weidner-Wells MA, Werblood HM, Goldschmidt R. Synthesis and Antibacterial Activity of novel oxazolidinone analogs containing substituted thiazole/ fused-bicyclic Gr. Chemical Research in Chinese Universities. 2006;**22**(4):459-464

[86] Vicini P, Geronikaki A, Anastasia K, Incerti M, Zani F. Synthesis and antimicrobial activity of novel 2-thiazolylimino-5-arylidene-4-thiazolid. Bioorganic & Medicinal Chemistry. 2006;**14**:3859

[87] Abdel-Wahab BF, Abdel-Aziz HA, Ahmed EM. Synthesis and antimicrobial evaluation of 1-(benzofuran-2-yl)-4-nitro-3-arylbutan-1-ones and 3-(benzofuran-2-yl)-4,5-dihydro-5-aryl-1-[4-(aryl)-1,3-thiazol-2-yl]-1H-pyrazoles. European Journal of Medicinal Chemistry. 2009;**44**:2632

[88] Mohd AS, Kanugala S, Ibrahim BS, Irfan K, Thipparapu G, Syed A, et al. Synthesis of new triazole fused imidazo[2,1-*b*]thiazole hybrids with emphasis on *Staphylococcus aureus* virulence factors. Bioorganic & Medicinal Chemistry Letters. 2019;**29**:126621

[89] Bharti SK, Nath G, Tilak R, Singh SK. Synthesis, antibacterial and anti-fungal activities of some novel Schiff bases containing 2,4-disubstituted thiazole ring. European Journal of Medicinal Chemistry. 2010;**45**:651-660

[90] Chimenti F, Bizzarri B, Maccioni E, Secci D, Bolasco A, Fioravanti R, et al. Synthesis and in vitro activity of 2-thiazolylhydrazone derivatives compared with the activity of clotrimazole against clinical isolates of *Candida* spp. Bioorg.Med.Chem. Lett. 2007;**17**:4635

[91] Geronikaki AA, Pitta EP, Liaras KS. Thiazoles and thiazolidinones as antioxidants. Current Medicinal Chemistry. 2013;**20**(36):4460-4480

[92] Karegoudar P, Karthikeyan MS, Prasad DJ, Mahalinga MB, Holla S, Kumari NS. Synthesis of some novel 2,4-disubstituted thiazoles as possible antimicrobial agents. European Journal of Medicinal Chemistry. 2008;**43**(2):261-267

[93] Hassan FA. Synthesis, characterization, anti-inflammatory, and antioxidant activities of some new thiazole derivatives. Current Journal of Applied Science. Technol. 2012;**2**(7):180-187

[94] Siddiqui N, Ahsan MW, Alam MS. Thiazoles: A valuable insight into the recent advances and biological activities. International Journal of Pharmacy and Pharmaceutical Sciences. 2009;**1**(3):136-143

[95] Estevão MS, Carvalho LC, Ribeiro D, Couto D, Freitas M, Gomes A, et al. Antioxidant activity of unexplored indole derivatives: Synthesis and screening. European Journal of Medicinal Chemistry. 2010;**45**:4869-4878

[96] Bozdag-Dundar O, Coban T, Ceylan-Unlusoy M, Ertan R. Radical scavenging capacities of some thiazolylthiazolidine-2,4-dione derivatives. Medicinal Chemistry Research. 2009;**18**:1

[97] Gouda MA, Berghot MA, Baz EA, Hamama WS. Synthesis and antioxidant evaluation of some new sulphadimidine incorporating thiophene moiety. European Journal of Chemistry. 2014;**5**(4):595-600

[98] Jaishree V, Ramdas N, Sachin J, Ramesh B. In vitro antioxidant properties of new thiazole derivatives. Journal of Saudi Chemical Society. 2012;**16**:371-376

Chapter 3

# Thiazolidinone-Related Heterocyclic Compounds as Potential Antitrypanosomal Agents

*Anna Kryshchyshyn, Danylo Kaminskyy, Philippe Grellier and Roman Lesyk*

## Abstract

Human African trypanosomiasis (HAT) and Chagas disease are neglected tropical diseases (NTDs) due to parasite protists from the *Trypanosoma* genus transmitted by insect vectors. Trypanosomiases affect mostly poor populations in the developing countries, and the development of new antitrypanosomal drugs is underinvested by governments and the pharmaceutical industry. In this chapter, we described the development of 4-thiazolidinone and thiazole derivatives with heterocyclic fragments which exhibit good inhibition of trypanosome growth and might constitute potential candidates for the development of new drugs against trypanosomiasis. Antitrypanosomal design, mainly within structure-based design, led to the synthesis of 5-ene-4-thiazolidinone-3-alkanecarboxylic acids; 2,3-disubstituted 4-thiazolidinones; thiazolidinone-pyrazoline, phenylindole-thiazolidinone, and imidazothiadiazole-thiazolidinone hybrids; as well as 4-thiazolidinone-based fused heterocycles, especially thiopyrano[2,3-*d*]thiazoles, and non-thiazolidinone compounds–namely, isothiocoumarine derivatives. Moreover, antitrypanosomal 4-thiazolidinones are of special interest in the search for new antimalarial and antileishmanial agents. Also many active anticancer agents among the abovementioned 4-thiazolidinones have been discovered.

**Keywords:** sleeping sickness, Chagas disease, antitrypanosomal drugs, thiazolidinone derivatives, hybrids

## 1. Introduction

Trypanosomatid infections belong to the neglected tropical diseases (NTDs)–a group of communicable diseases spread in 149 countries in the tropical and subtropical regions of the globe and affecting more than 1 billion people [1]. These vector-borne parasitic diseases are associated with poverty, contact with infectious vectors, as well as limited accesses to health services [2]. Human trypanosomiasis is caused by kinetoplastids, flagellated protists of *Trypanosoma* genus transmitted by an insect vector [3].

*Trypanosoma brucei gambiense* (*T.b. gambiense*) and *Trypanosoma brucei rhodesiense* (*T.b. rhodesiense*) are transmitted by the tsetse fly and cause two forms of human African trypanosomiasis (HAT) known as sleeping sickness when neurological manifestations associated with presence of parasites in the brain become apparent [4]. *T.b. gambiense* accounts for more than 98% of reported cases; *T.b. rhodesiense* is responsible for an acute infection and represents under 2% of reported cases [5]. Other *Trypanosoma* species (e.g., *T. vivax*, *T. congolense*, and *T. evansi*) affect cattle, causing animal African trypanosomiasis (Nagana) and contributing to livestock losses. Cattle are also a reservoir of infection for human trypanosomes [6]. Therefore, the necessity to control animal trypanosomiasis should not be underestimated within the concept of "one health" [7]. There had been several devastating HAT epidemics during the twentieth century, the last one occurred in the late 1990s with estimated near 300,000 cases. Thanks to the coordinated work of the WHO and governmental and nongovernmental organizations to combat NTDs, the number of cases reported in 2009 has dropped below 10,000 for the first time in 50 years. This trend persists, and in 2019 there were less than 1000 incidences of HAT, although the estimated number of people being at risk of infection is near 65 million. First signs and symptoms of HAT are observed a few weeks after infection. During the first hemolymphatic stage, trypanosomes invade the human host and locally multiply spreading via the lymph and blood to various peripheral organs. The following meningoencephalitic stage develops when the parasites invade the brain parenchyma crossing the blood-brain barrier. The second stage of HAT is characterized by neurological disturbances and neuropsychiatric and sleep disorders [4]. If left untreated, the disease leads to coma and death [8]. Vector control is an important issue in the efforts taken to eliminate HAT. This is evidenced by the elimination of trypanosomiasis in Zanzibar Island due to tsetse clearance. This approach is still difficult to implement on a continent; therefore chemotherapy remains the main tool in the HAT management [9]. Difficulty of vaccine development because of the antigenic variation of the parasite surface proteins has been one more unsolved problem [10].

Chagas disease (American trypanosomiasis) caused by *Trypanosoma cruzi* (*T. cruzi*) is a devastating human disease with about 8 million infected people mostly in Latin and South America. Over the past decades, due to migration and population mobility, Chagas disease cases were reported in Europe, the United States, and Canada [11]. It is transmitted to man during the bite of a bloodsucking triatomine bug, via its feces or urine through skin breaks or mucous membranes, and occasionally causing outbreaks through contaminated food. Transmission through blood transfusion and pregnancy is also possible and, less frequently, through organ transplantation or laboratory accidents [12–14]. Once the parasite reaches the human host, it multiplies in the host's cells in the amastigote form that differentiates into the infective trypomastigote form, which is released after the host cell rupture, causing inflammatory reactions and leading to megaesophagus, megacolon, and cardiac conduction disturbances [15, 16]. Since Chagas disease was discovered in 1909, numerous studies have been carried out to investigate the pathogenesis of acute and chronic phases of the disease [11]. While the acute phase is often asymptomatic or characterized by non-specific symptoms, except sometimes occurring chagoma or Romaña sign, the chronic phase can be subdivided into an asymptomatic indeterminate phase and a symptomatic determinant phase [17]. Between 60% and 70% of serologically positive patients have no manifestation of the disease; in the remaining 30–40%, cardiac and gastrointestinal complications develop, indicating a symptomatic determinant phase [18]. If earlier autoimmune reactions were thought to be the primary factors leading to the lesions associated with the chronic stage, recent investigations showed that the persistence of parasites also contribute to the inflammatory processes, leading to cardiac or gastrointestinal complications. Therefore treatment success depends greatly on the elimination of *T. cruzi* from the organism [16].

## 1.1 Treatments of trypanosomiasis

### 1.1.1 HAT

Suramin, pentamidine, melarsoprol, and eflornithine have been used to treat HAT for decades [19, 20] (**Figure 1**). An important advance was the development of the nifurtimox-eflornithine combination therapy (NECT), which has now became the standard first-line treatment for the second stage of *T.b. gambiense* HAT [20, 21]. Choice of the drug as well as duration of treatment depends on the stage of the disease and the parasite subspecies. Pentamidine isethionate is the first-line treatment for the first stage of *T.b. gambiense* disease, while suramin is used in the treatment of first stage of HAT caused by *T.b. rhodesiense.* Intravenous treatment with suramin, although usually effective, especially when given early in the disease, can result in potential complications such as renal failure, skin lesions, anaphylactic shock, bone marrow toxicity, and neurological complications. Pentamidine, administered by the intramuscular route or intravenously, despite non-negligible undesirable effects (hypoglycemia, prolongation of the QT interval on electrocardiogram, hypotension, and gastrointestinal features), is in general well tolerated by patients and is usually effective [22, 23]. NECT, being the first-line treatment for the second stage of *T.b. gambiense* disease, consists of nifurtimox delivered orally and eflornithine delivered intravenously. In the case of contraindications to nifurtimox, eflornithine may be given as a monotherapy for *T.b. gambiense* HAT (meningoencephalitic stage), but it is not recommended for *T.b. rhodesiense* disease [4, 24]. Melarsoprol is restricted to the treatment of the second stage of *T.b. rhodesiense* HAT because of severe adverse drug reactions, such as an encephalopathy syndrome that occurs in 5–18% of all treated cases and may be fatal [25–27]. The only indication of

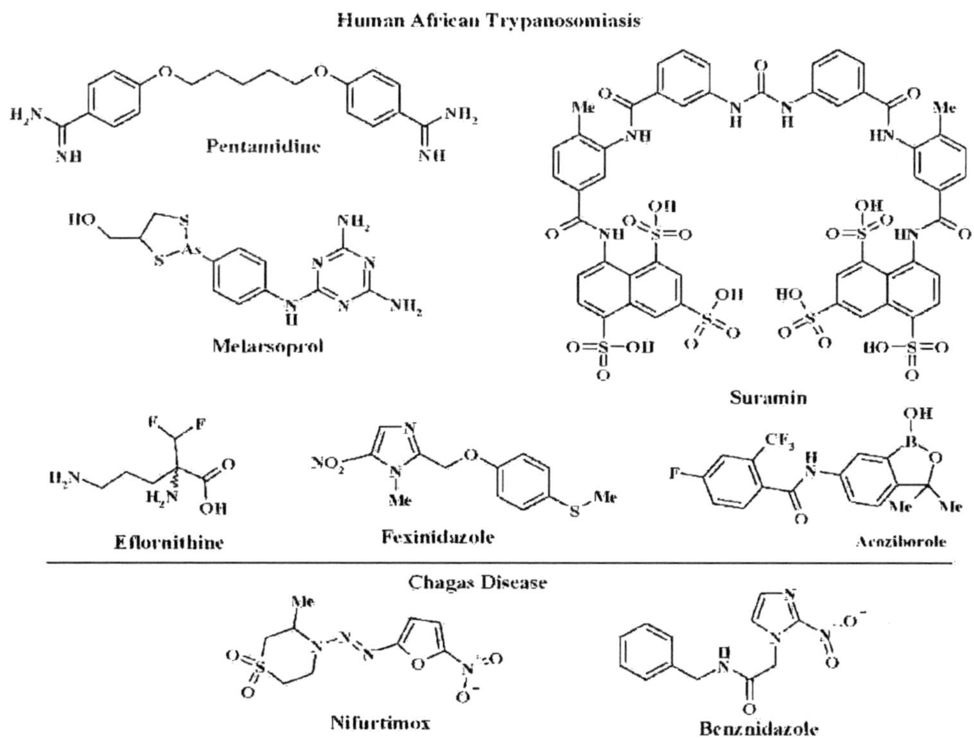


**Figure 1.**
*Drugs used for human trypanosomiasis treatment.*

melarsoprol for the treatment of *T.b. gambiense HAT* appears in the case of disease relapse after administering NECT or eflornithine monotherapy.

New effective oral monotherapy of HAT with fexinidazole has been developed and approved, so in 2018 the European Medicines Agency's (EMA) Committee for Medicinal Products for Human Use issued a positive opinion for fexinidazole treatment of *T.b. gambiense* HAT [28–30]. According to new WHO guidelines, under particular conditions, fexinidazole may replace pentamidine as first-line treatment in patients with the first stage of *T.b. gambiense* HAT and replace nifurtimox-eflornithine combination therapy as first-line treatment in patients with the second stage of *T.b. gambiense* HAT with fewer than 100 cerebrospinal fluid white blood cells per μL. These recommendations cannot be applied for the treatment of patients younger than 6 years or with a bodyweight less than 20 kg [31, 32]. One more new oral compound developed for treatment of all stages of *T.b. gambiense* HAT is acoziborole being at late Phase II/III of clinical trials [31].

#### *1.1.2 Chagas disease*

Only two drugs are currently available, nifurtimox and benznidazole (**Figure 1**), that both are active in the acute stage of the disease (up to 80% efficacy), though of limited efficacy against the established chronic stage of the disease [14, 33]. Benznidazole is a nitroimidazole, which generates radical species in aerobic and anaerobic conditions [34], and is the agent of choice for monotherapy of Chagas disease because of its extensive security and efficacy profile. Generalized adverse effects [17] as well as occasionally reported resistance to benznidazole make nifurtimox usage an alternative treatment. Both drugs produce important adverse reactions, especially in adults, because newborn, nursing, and small children tolerate these drugs better [35, 36]. In the acquired acute period, 70% of the cases are cured, and in newborn and nursing children with congenital Chagas disease, 98–100% cure is obtained. On the one hand, there is evidence about efficiency of benznidazole in early chronic infections [33], but on the other hand, the expediency of antitrypanosomal treatment in the chronic stages remains controversial, because of significant toxicity profiles and the unproven role in preventing the cardiomyopathy progression. Therefore, therapy for the majority of patients suffering from chronic Chagas disease consists mostly in nonetiologic treatments [11]. New effective and safety drugs are needed, especially for the chronic stage treatment [37].

### 1.2 Drug discovery strategies

As American and African trypanosomiases affect mostly poor population in the low- and middle-income countries and have not been interesting for the big pharmaceutical companies for years, a number of public and private institutions, partnerships, and consortia were initiated. For example, the Special Programme for Research and Training in Tropical Diseases of WHO (WHO/TDR), the European Commission [38] as a government agency, or the international Drugs for Neglected Diseases Initiative (DNDi) [39] had emerged. The work of these organizations has had an undeniable positive impact on the development of novel therapies and for the elimination of trypanosomiasis.

In general, three known major approaches to novel drug development, including antitrypanosomals, may be outlined: (i) ligand-based approach, (ii) target-based drug discovery [40], and (iii) phenotype-based drug discovery [41]. Different types of compounds, namely, thiosemicarbazones, thiazolidines, triazole- and furan-based compounds, benzofuran derivatives, peptidyl compounds, peptidomimetics acyl- and arylhydrazones, etc. have been studied as novel antitrypanosomal agents [42].

In ligand-based approaches, already known active synthetic and natural compounds or approved drugs are used as starting scaffolds to develop novel agents [43]. For example, development of pentamidine analogues resulted in the lead compound DB289 that underwent preclinical and clinical studies [44]. Other examples of the abovementioned approach are label extension or search for the new indications of existing drugs [45].

Target-based approaches involve screening of drug libraries with established targets, within target repurposing strategy, or screening libraries of novel compounds against a definite protein target. The structures of identified hit compounds are often optimized in order to increase their selectivity and pharmacokinetic properties or decrease their toxicity [46]. It should be mentioned that the target validation status used in the antitrypanosomal drug discovery often has not been clear. WHO/TDR Target Prioritization Network helps the scientists in the rational drug design of antiparasitic agents including antitrypanosomal drugs. The TDR Targets database, developed by this organization, contains information on validated, essential, as well as putative targets; it also can serve as a tool for prioritization of targets in whole genomes [47, 48].

### 1.3 Examples of targets used in novel antitrypanosomal agent development

#### *1.3.1 Trypanosomatid peptidases*

Numerous studies showed that intra- and/or extracellular trypanosomatid peptidases play important roles in different cell functions including invasion, intracellular survival, replication, differentiation, infectivity, immune evasion, and nutrition. "Validated" trypanosomatid peptidases belong to the endopeptidases and include cruzipain, prolyl oligopeptidases (POPs; *T. cruzi*), congopain (*T. congolense*), rhodesain (*T.b. rhodesiense*), and brucipain (*T.b. brucei*) [49]. For example, the cysteine peptidase cruzipain being differentially expressed in the different stages of *T. cruzi*, along with other peptidases, is responsible for parasite survival, differentiation, and growth. Cruzipain is a sulfated glycoprotein, which is investigated not only as a drug target but also as a candidate for vaccine development [50]. Selective inhibitors of this peptidase arrest metacyclogenesis in vitro and block the proliferation of both extracellular epimastigotes and intracellular amastigotes. The main lysosomal cysteine peptidases *rhodesain*, *brucipain*, and *congopain* are cathepsin L-like proteases [49]. They may play a role in anemia and immunosuppression due to infection, and conversely, anti-cysteine peptidase antibodies may modulate the trypanosome-induced pathology [51]. Oligopeptidases B and Tc80 are serine protease representatives of the prolyl oligopeptidase family [49]. Oligopeptidase B is involved in the mammalian host cell invasion by the trypomastigotes [52]. It retains full catalytic activity when released into the host bloodstream providing anomalous degradation of host peptide hormones that reinforces the importance of its protein-processing activity [53]. POP Tc80 has been detected in all the developmental stages of *T. cruzi* but is secreted by the trypomastigotes. POP Tc80 was shown to exhibit the unusual property of cleaving collagens I and IV, fibronectin, and peptide hormones. POP TC80 inhibitors block the host cell invasion by trypomastigotes; selectivity between parasitic and human POPs toward inhibitors could be expected [54, 55].

#### *1.3.2 Nitroreductases*

Nitroreductases are mainly associated with the nifurtimox mode of action. The activity of type I nitroreductase is believed to be "oxygen-insensitive" as it does not involve oxygen in the reduction process and therefore does not cause the reactive oxygen species production. In contrast, the activity of type II nitroreductase results in the production of superoxide anions, so it is considered "oxygen-sensitive." Nifurtimox selectivity toward parasites was associated with the expression of type I nitroreductase. But, considering that nifurtimox-treated trypanosome extracts contain superoxide anions and nitro anion radicals, an oxidative stress with a type II nitroreductase involving is generally accepted to be the main trypanocidal mode of its action [56].

### *1.3.3 Dolicholphosphate mannose synthase*

Dolicholphosphate mannose synthase is a mannosyltransferase critically involved in glycoconjugate biosynthesis in *T. brucei*. Variant surface glycoprotein (VSG) dimers, covering the surface of the parasite and undergoing constant antigenic variation, act as a physical diffusion barrier for components of the innate immune system as the parasite switches between many immunologically distinct VSG genes. All VSG variants are linked to the plasma membrane via glycosylphosphatidylinositol (GPI) anchors. The biosynthesis of GPI anchor was shown to be essential for viability of the bloodstream form of *T. brucei*, thus validating it as a drug target against HAT [57].

### *1.3.4 Dihydrofolate reductase*

Dihydrofolate reductase (DHFR) is a key enzyme of the folate metabolism, deeply studied in the design of a number of anticancer, antibacterial, and antimalarial agents [58]. Detailed structural analysis of *T. brucei* and *T. cruzi* DHFRs showed their differences from the human enzyme, indicating them as attractive targets for the development of selective antitrypanosomals. Well-known DHFR inhibitors, as trimethoprim and pyrimethamine, are weakly active against *T. brucei* and *T. cruzi* DHFR unlike methotrexate being reported to inhibit *T. cruzi* enzyme in nanomolar concentrations [59].

### *1.3.5 Trypanothione reductase*

Trypanothione reductase (TryR)–an enzyme of the NADPH-dependent flavoprotein oxidoreductase family–converts trypanothione disulfide into the physiologically relevant reduced dithiol. TryR is essential for growth of trypanosomatids as in the absence of catalase and glutathione peroxidase, the trypanothione system is involved in response to an oxidative stress. To some extent, trypanothione disulfide serves as glutathione in mammalian cells. Although mammalian glutathione reductase is homologous to parasite TryR, there are significant differences in their active sites [60, 61].

### *1.3.6 Kinases*

The genomic analysis of *T. brucei* and *T. cruzi* revealed 156 and 171 eukaryotic protein kinases (PKs) in the parasite genomes. Atypical PKs representing four families, RIO, alpha, PIKK, and PDK, had also been discovered. Such an amount of PKs that are key mediators of signal transduction indicates the important role they play in trypanosomatid life cycles [62]. The differences in structure between trypanosomatid PKs and mammalian PKs as well as the evidence that some trypanosomatid PKs are vital for the parasite make these enzymes suitable for the antitrypanosomal drug search [63].

#### *1.3.7 Triosephosphate isomerase*

Triosephosphate isomerase (TIM) catalyzes the interconversion between glyceraldehyde 3-phosphate and dihydroxyacetone phosphate in the glycolytic pathway [64]. The presence of TIM in both human and parasite (68–74% of identity between both enzymes) makes targeting this enzyme problematic [65]. The structures of *T. brucei* and *T. cruzi* TIMs are also quite similar, except the structural differences that influence their different sensitivity to sulfhydryl reagents. *T. cruzi* TIM showed the highest sensitivity, constituting a good target for the development of selective therapeutics for the Chagas disease [66].

#### *1.3.8 Farnesyl diphosphate synthase*

Farnesyl diphosphate synthase (FPPS) catalyzes isopentenyl diphosphate and dimethylallyl diphosphate condensation resulting in the formation of geranyl diphosphate and subsequently farnesyl diphosphate that are precursors for the biosynthesis of isoprenoid derivatives (e.g., dolichols, sterols) and for protein prenylation. Bisphosphonates, such as alendronate and risedronate, are considered to be ligands for *T. cruzi* FPPS [67]. FPPS is an attractive target for antichagasic drug development as it is essential for parasite's growth and proliferation [68, 69].

#### *1.3.9 Cyclic nucleotide-specific phosphodiesterases*

Cyclic nucleotide-specific phosphodiesterases (PDEs) are also shown to be promising antitrypanosomal drug targets [70]. There are four distinct PDE families encoded in the genome of *T. brucei* [71].

Kinase inhibitors [72], such as human *Aurora kinase* inhibitors, typified by danusertib [73], and human *epidermal growth factor receptor* (EGFR) inhibitors lapatinib and canertinib [74] are examples of successful implementations of the target repurposing strategy when pathogen targets are matched with known homologous human targets.

One more variation of target-based drug design is the screening of known drug libraries in order to establish new pharmacological profile. For example, screening of a library of bioactive compounds against TryR [75] led to identification of a new class of TryR inhibitors based on indatraline, a nonselective monoamine reuptake inhibitor [76].

#### *1.3.10 Lanosterol 14α-demethylase*

Lanosterol 14α-demethylase or CYP51, which belongs to the family of cytochrome P450s, is one of the most promising antitrypanosomal targets. This enzyme is involved in the ergosterol biosynthesis, taking part in the production of components of the plasma membranes and serving as precursors for regulatory molecules that modulate growth, division, differentiation, and development processes [77, 78]. Fungicides as well as clinically used antifungal azoles inhibit CYP51 that along with the resemblance of sterol biosynthesis in trypanosomatids to such in fungi [79], makes *lanosterol 14α-demethylase* an attractive target for the design of antitrypanosomal agents.

In the era of target therapy, phenotypic screening that lies in pharmacological screening of chemical libraries against whole-cell or biological system should not be neglected [80–82]. This approach is particularly advantageous in the search of antitrypanosomals [83, 84], as the success strongly depends on the penetration properties of the drug into the parasite as well as on the crossing of the blood-brain barrier.

Sometimes, high-affinity ligands toward validated trypanosomal targets were shown ineffective *in vivo* against the parasite because not crossing the membranes, that is one more argument in favor of the whole-cell phenotypic assays [42]. Target resolution from phenotypic hits may also contribute to drug discovery process [84].

It should be mentioned that the parasites of *Leishmania* genus belong to the same order *Kinetoplastida* as *Trypanosoma* ssp. sharing some phylogenetic similarities [85]. Similar structural and biochemical features include, for example, special organelles (kinetoplast (mitochondrion with a discrete structured DNA body), glycosomes (involved in glycolysis)), a sub-pellicular microtubular corset, and a unique thiol metabolism [10, 86]. Interesting is that hit compounds found in antitrypanosomal screening may be used for the design of agents against *Leishmania* ssp. [87, 88] or *vice versa.*

## 2. 4-Thiazolidinone frame in the design of antitrypanosomals

4-Thiazolidinones are well-known class of azoles, which have been investigated for many decades as useful tools for the design and development of new drugs [89–93]. 4-Thiazolidinone scaffolds (2,4-thiazolidinedione, rhodanine (2-thioxo-4-thiazolidinone), 2-alkyl(aryl)-substituted and 2-amino(imino)-substituted 4-thiazolidinones) (**Figure 2**) are used as privileged structures and substructures in the modern medicinal chemistry [94–98] for the design of new anti-inflammatory, antitumor, antimicrobial, antidiabetic, antibacterial agents, etc. The synthetic approaches for these heterocycles are well known and described [96].

Majority of the 4-thiazolidinone-based hit and lead compounds, drug-like molecules, and approved drugs belong to derivatives containing the exocyclic double bond at C5 position—5-ene-4-thiazolidinones [96, 97]. These compounds, especially rhodanine derivatives, are possible Michael acceptors and are claimed as frequent hitters or pan-assay interference compounds (PAINS), being treated as useless in the drug discovery process because of their possible/predicted insufficient selectivity [99]. This statement should not be regarded as a general knockout criterion that excludes such screening hits from further development and should be studied in more detail [96, 97, 100, 101]. Therefore, "4-thiazolidinones and related scaffolds should not be regarded as problematic or promiscuous binders *per se*" [95], while "positive" properties of Michael acceptors should be effectively used [95, 97]. For instance, Michael acceptors are among the most effective activators of Nrf2 through the Keap1 modification, which open new perspectives in the treatment of inflammation, cancer, etc. [102]. Moreover, Michael acceptor properties are often not confirmed in experimental studies [103, 104] under conditions similar to physiological ones.

The search for new antimicrobial and antiparasitic agents based on 4-thiazolidinone cores is one of the earliest directions of biological studies of 4-thiazolidinones. The structural similarity of 4-azolidinones with penicillin antibiotics was the stimulus to the study of such type of activity [90, 105–107]. However,

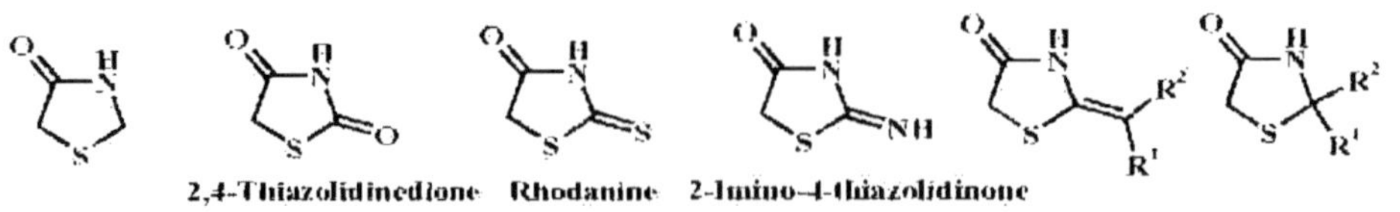


**Figure 2.**
*Main 4-thiazolidinone-based scaffolds.*

currently the effects of 4-thiazolidinones are not related to the "penicillin" mode of action [91, 96].

In the field of antiprotozoal agent search, the design of antitrypanosomal agents based on thiazolidinone scaffolds is of special interest [42, 108]. Data on the search for new antitrypanosomal agents among 4-thiazolidinone derivatives present mostly investigations on the inhibition of parasite growth (phenotype screening) mainly within a privileged substructure-based design. A much smaller number of publications are devoted to the study of the mechanism of action or the design of high-affinity ligands to "validated" targets [42, 96].

One of the arguments for the study of 4-thiazolidine-based compounds as antitrypanosomal agents is the thesis that thiazoles, especially 4-thiazolidinones, are considered as thioureas/thiosemicarbazones' cyclic analogues and biomimetics [42, 96, 108, 109]. Different (thio)ureas/(thio)semicarbazides were reported as inhibitors of the trypanosome proliferation [110–112] and had shown high affinity to the antitrypanosomal targets: cruzain and rhodesain [109, 113], cysteine proteases [114], etc. Different classes of "drug-like" molecules based on a thiazolidinone scaffold have been designed and synthesized in the process of search for antitrypanosomals [42, 115–119]. One of the most prominent directions is the conjugation of the thiazolidinone core with other different molecular fragments (mainly privileged substructures) [120, 121] that proves the efficiency of a molecular hybridization approach and a hybrid pharmacophore approach for the design of new antitrypanosomals [122–124].

Combination of 4-thiazolidinone and pyrazoline cores led to the synthesis of rows of promising trypanocidal agents **(1–4)** (**Figure 3**) with sub-micromolar activity levels against *T.b. brucei* and *T.b. gambiense* [121, 125–127] and low toxicity levels against mammalian cells.

Compounds with an enamine linker **5**, **6** (**Figure 4**) were designed based on the early hits **1**, **2** (4-thiazolidinone and pyrazoline cores are bonded without additional linker). Most active compounds from these series, 5-[5-(4-methoxyphenyl)-3-naphthalen-2-yl-4,5-dihydropyrazol-1-ylmethylene]-3-methyl-2-thioxothiazolidin-4-one ($IC_{50}$ = 0.6μM) and 5-[5-(2-hydroxyphenyl)-3-(4-methoxyphenyl)-4,5-dihydropyrazol-1-ylmethylene]-3-(3-acetoxyphenyl)-2-thioxothiazolidin-4-one ($IC_{50}$ = 0.7μM), possess sub-micromolar activities and high selectivity indexes [121]. Elongation of the enamine bearing linker group (compounds **6**) led to a decrease of the activity, and modification of the N3 position of thiazolidinone core (compounds

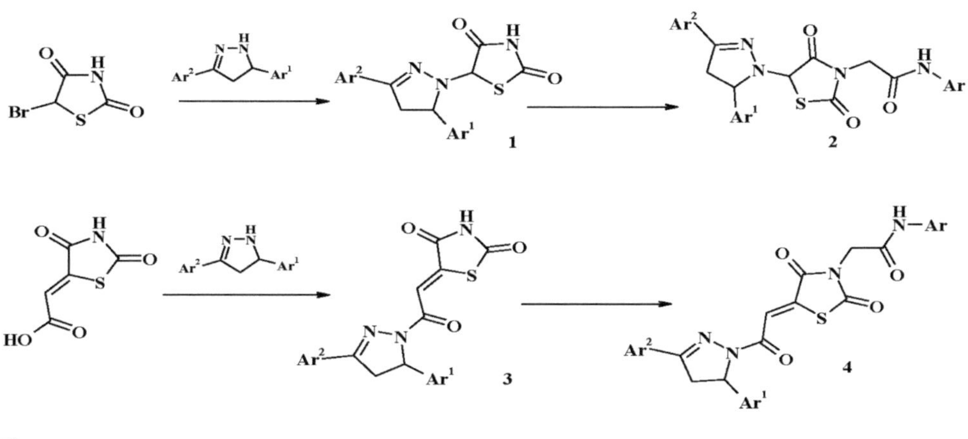

**Figure 3.**
*Thiazolidinone-pyrazoline conjugate synthesis.*

**5** as well as compounds **2** and **4**) was considered as crucial for the trypanocidal activity (methyl or small aryl fragments are desirable) [127].

It should be noted that mentioned compounds are considered as prominent anticancer agents [127] and compounds **5** showed a strong antileukemic activity with an apoptotic-related mitochondria-dependent mode of action with a prooxidant action [128].

Related 4-thiazolidinone-pyrazoline conjugates 7 (**Figure 5**) synthesized based on an isorhodanine (4-thioxo-2-thiazolidinone) core [129, 130] were also studied in vitro against *T.b. brucei*, and compounds with a micromolar activity were identified [126].

A moderate antitrypanosomal activity of pyrimidine-thiazolidine-4-one hybrids **8** (**Figure 6**) was reported against bloodstream forms of *T.b. brucei* ($IC_{50}$ = 25–100 μM) [131].

Related 2,3-substituted 4-thiazolidinones **9** with simple aromatic substituents at the position C2 and N3 also possessed low to moderate levels of activity against *T.b. brucei* and *T.b. gambiense* [132]. The synthetic methods for their obtaining are based on the one-pot three-component reaction of amine, oxocompound, and thioglycolic acid or its derivatives [133, 134]. It should be noted that the abovementioned derivatives of thioglycolic acids, namely, 2-mercaptoacrylic acids, can be easily synthesized or formed via a metabolic transformation based on simple 5-arylidenerhodanines (**Figure 7**) and possess similar pharmacological profiles [135].

Moreover, simple 5-ene-2,4-thiazolidinones were proposed as possible scaffolds for the design of new antitrypanosomal agents as pteridine reductase 1 inhibitors [136].

5-Arylidenerhodanine-3-acetic acids **10** (**Figure 8**) as one of the most studied types of thiazolidinones were reported to inhibit the activity of the dolicholphosphate mannose synthase and the GPI anchor synthesis and exhibited trypanocidal activity against the bloodstream forms of *T.b. brucei* ($ED_{50}$ = from 96 to 492 μM) [57]. Structure optimization of 4-thiazolidinone-carboxylic acids, including compounds with anticancer properties [137, 138], allowed to obtain a series of 2-(5-aminome thylene-4-oxo-2-thioxothiazolidin-3-yl)-3-phenylpropionic acid ethyl esters **11**. Among them, several hit compounds (2-{5-[(5-chloro-2-methoxyphenylamino)-methylene]-4-oxo-2-thioxothiazolidin-3-yl}-3-phenylpropionic acid ethyl ester, 2-(5-{[2-methyl-5-(morpholine-4-sulfonyl)phenylamino]-methylene}-4-oxo-2-thioxothiazolidin-3-yl)-3-phenylpropionic acid ethyl ester, and 4-{[3-(1-ethoxycarbonyl-2-phenylethyl)-4-oxo-2-thioxothiazolidin-5-ylidenemethyl]-amino}-benzoic

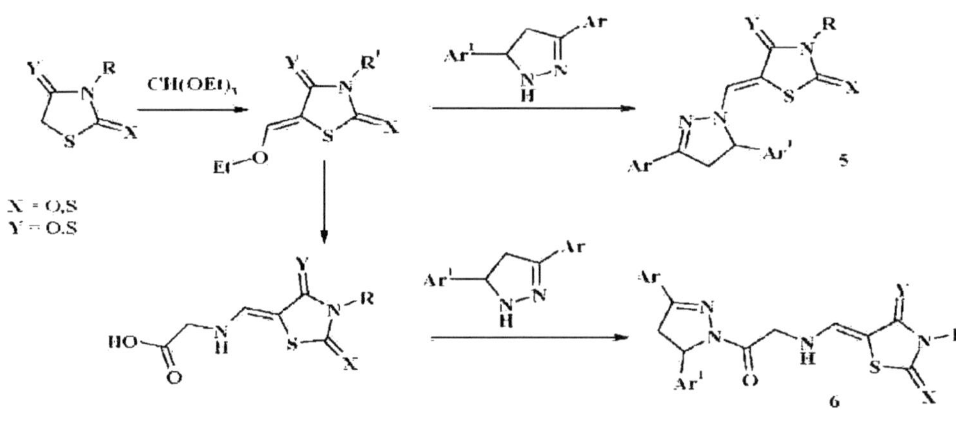


**Figure 4.**
*5-Enamine 4-thiazolidinone-pyrazoline conjugates.*

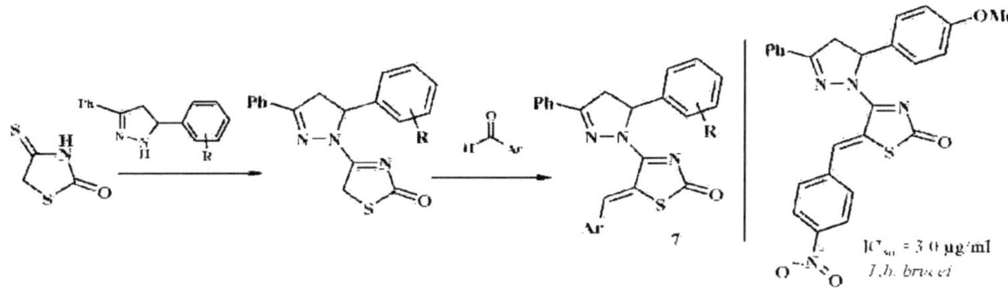

**Figure 5.**
*4-Substituted 2-thiazolidinone synthesis.*

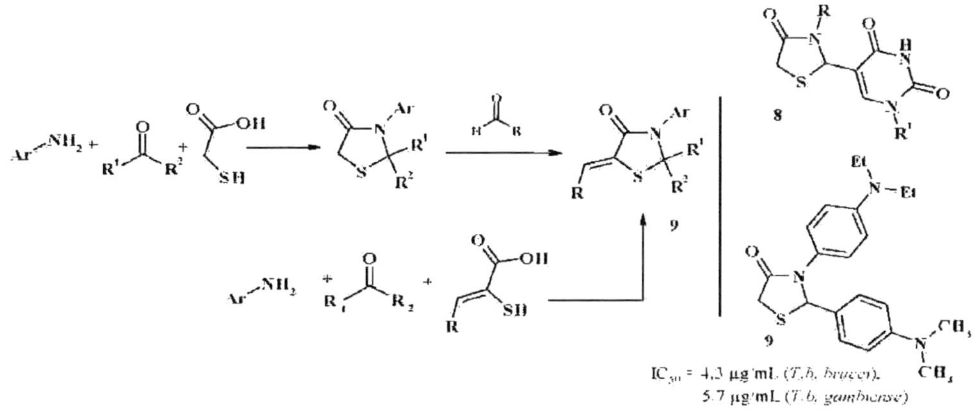

**Figure 6.**
*2,3-Disubstituted 4-thiazolidinone synthesis.*

acid ethyl ester) inhibited the in vitro growth of *T.b. brucei* and *T.b. gambiense* at nano- and sub-micromolar concentrations ($IC_{50}$ = 0.027–1.936 μM), and significant selectivity indices (SI = 108–1396) were calculated [139].

Screening of a focused kinase inhibitor library against cultures of *T.b. brucei* allowed identifying a series of active compounds based on 2,4-diaminothiazoles, some of them possessing antitrypanosomal activity at the nanomolar range [140]. Combination of thiazolidine scaffold with a thiophene moiety yielded thiophen-2-iminothiazolidine hybrids that showed trypanocidal activity in vitro against *T. cruzi* (amastigote and trypomastigote forms) and cruzain inhibition activity [115].

One of the directions for the design of new antitrypanosomal agents using a molecular hybridization approach is the utilization of hydrazone fragments (**Figure 9**) as the linker group for the connection of the thiazole/4-thiazolidinone scaffold with the other molecular fragments [117, 141–147].

Screening of 4-thiazolidinone-hydrazones against *T. cruzi* yielded active and non-cytotoxic compounds **12** (**Figure 10**) [148, 149]. The 2-hydrazolyl-4-thiazolidinone-5-carboxylic acid derivatives **13** have shown promising activity on the cruzipain protease. Compounds were selected based on a virtual screening of 500,000 chemical structures (ZINC5 database). Structurally related compounds **14** (with exocyclic double bond at C5 position) showed the highest antiproliferative activity when screened on *T. cruzi* epimastigotes but were inactive toward cruzipain [127]. 5-Alkyl-4-thiazolidinone-2-hydrazones **15** tested in a cruzain inhibition assay and against cultures of the epimastigote and trypomastigote forms (*T. cruzi*, Y strain) inhibited the cruzain activity and showed an antiproliferative activity

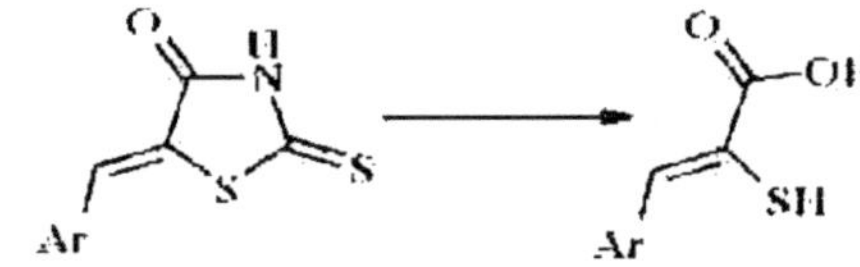

**Figure 7.**
*Thiazolidinone-based approach to 2-mercaptoacrylic acid formation.*

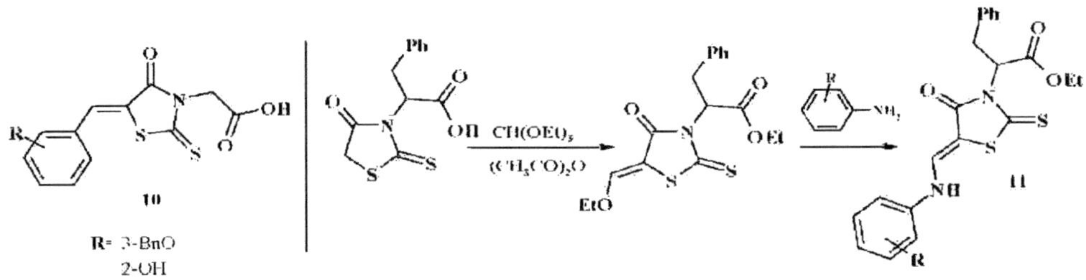

**Figure 8.**
*5-Ene-4-thiazolidinone-3-carboxylic acid synthesis.*

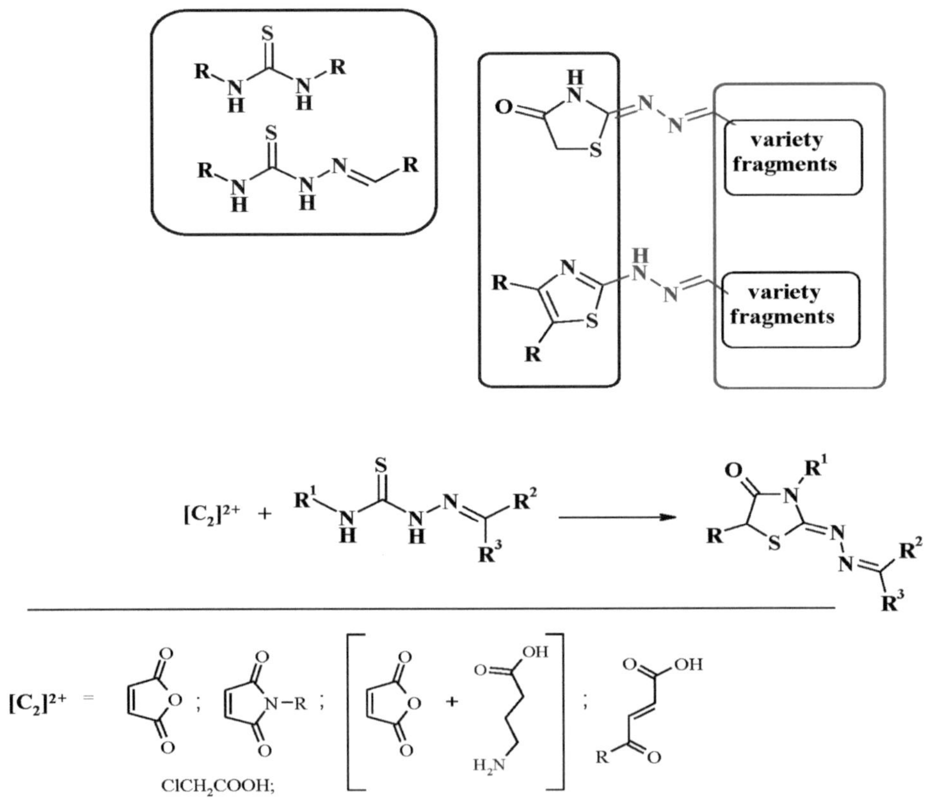


**Figure 9.**
*2-Hydrazono-4-thiazolidinone synthesis.*

at non-cytotoxic concentrations [150]. Study of analogues, namely, 2-imino-1,3-thiazoles, showed that the bioisosteric replacement of thiazolidine cycle with thiazole led to loss of the cruzain inhibitory activity and a significant reduction of the trypanocidal activity. The most potent cruzain inhibitor 2-((1-phenoxypropan-2-ylidene)hydrazono)-3-phenyl-5-isopropylthiazolidine-4-one also impaired intracellular trypomastigote development and attenuated trypomastigote invasion

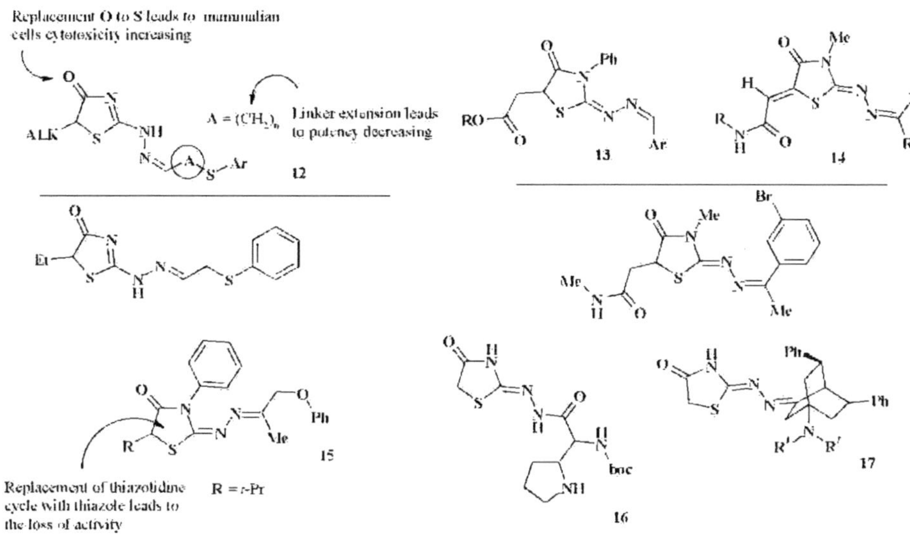


**Figure 10.**
*4-Thiazolidinone-hydrazones as trypanocidal agents.*

of macrophages; however it did not eradicate parasite in mice [150]. 2-Aminoacyl-4-thiazolidinone derivatives also showed good trypanocidal properties against *T. cruzi*; the proline derivative **16** showed differences of efficiency according to the parasite strains tested (Y strain vs Colombian strain). Docking analysis to *T. cruzi* cruzain that corroborated the experimental $IC_{50}$ data and analysis of the binding characteristics of tested ligands revealed important interactions, which explain the affinity of such derivatives to cruzain [42, 151]. Combination of 4-dialkylaminobicyclo[2.2.2]octane fragment with the 5-unsubstituted 4-thiazolidinone core led to compounds **17** with weak to moderate activity against *T.b. rhodesiense* [152].

Molecular hybridization of the thiazole ring with a pyridine moiety through a hydrazine bridge led to identification of selective *N*-[3-phenyl-3*H*-thiazol-2-ylidene]-*N′*-(1-pyridin-2-yl-ethylidene)-hydrazines inducing the parasite death via an apoptotic mechanism [153]. Combination of a thiazole core with fused [6+5] or [6+6] scaffolds turned out to be especially interesting, leading to highly active and selective antitrypanosomal agents. Synthesized indanone-thiazole hybrids **18** (**Figure 11**) provide good trypanocidal properties against *T. cruzi* ($IC_{50}$ within 0.09–1.35 μM, Tulahuen 2 strain); these compounds were also characterized by low mammalian cell cytotoxicity [154].

Development of optimization directions of thiazolidinone-hydrazone structures led to new hybrid molecules bearing thiazolidinone/thiazole and 2-phenylindole/6-phenyl-imidazo[2,1-*b*][1, 3, 4]thiadiazole cores with hydrazone linkers **19**, **20** [155]. Compounds with sub-micromolar levels of trypanocidal activity toward bloodstream forms of *T.b. brucei* and *T.b. gambiense* and relatively low cytotoxicity upon human primary fibroblasts were identified, as well as some aspects of SAR (**Figure 12**) were derived.

Compounds with a 2-arylindole fragment were more active than 6-aryl-imidazo[2,1-*b*][1, 3, 4]thiadiazole analogues. For the compounds without phenyl ring attached to the indole fragment, no significant antitrypanosomal activity was found as well as for the compounds with a C5-ene-fragment in the 4-thiazolidinone core [155].

The main features of the molecular structure of thiazolidinone-hydrazone-based compounds can be outlined as the following: (i) thiazole core (position C4, small aryl or alkyl substituent; C5 position, unsubstituted or small alkyl fragment;

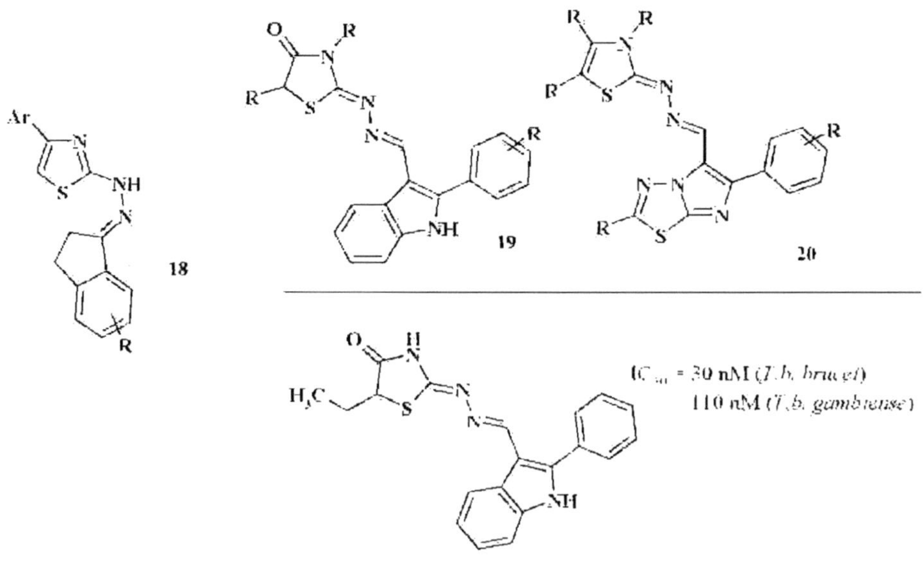


**Figure 11.**
*Thiazolidinone-indanone/indole/imidazothiadiazole hybrids.*

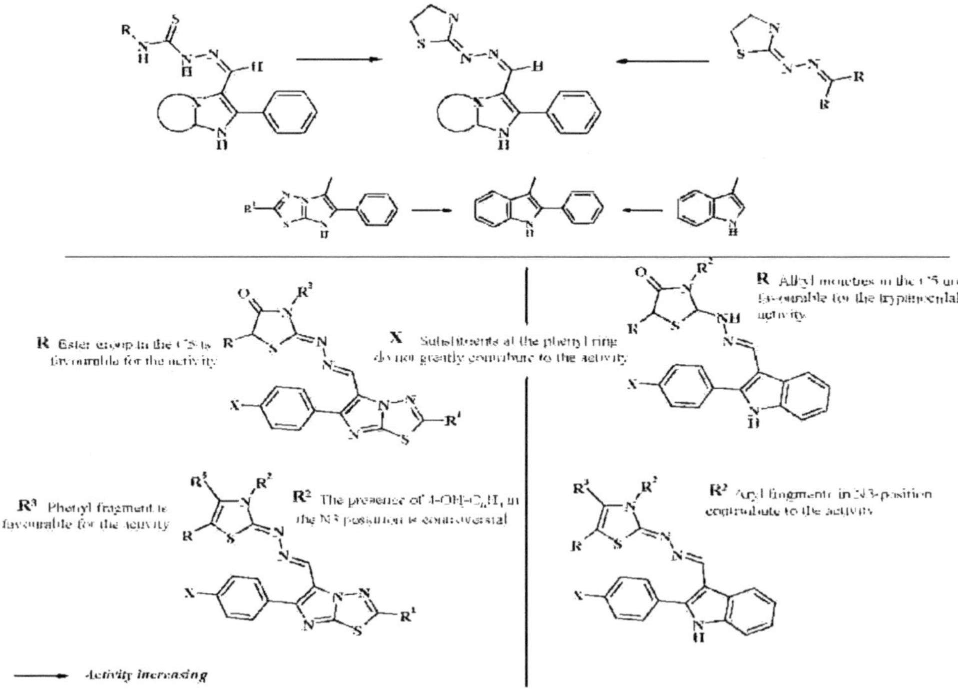


**Figure 12.**
*SAR of indole/imidazothiadiazole-thiazolidinone/thiazole hybrids.*

N3 position, variety of substituents) or 4-thiazolidinone core (C5 position, unsubstituted or small alkyl fragment); (ii) hydrazone linker in the C2 position of the main core; (iii) additional molecular fragment, diverse substituents (from simple alkyl(aryl)ydene fragment to privileged heterocyclic cores); and (iv) target compounds imitating the thiosemicarbazones with trypanocidal activity [147, 153, 155].

The "fixation" of the hydrazone fragment in a pyrazoline core (**Figure 13**) as one of the methods of such compound optimization has been also described for the synthesis of active compounds **21, 22** [126, 127].

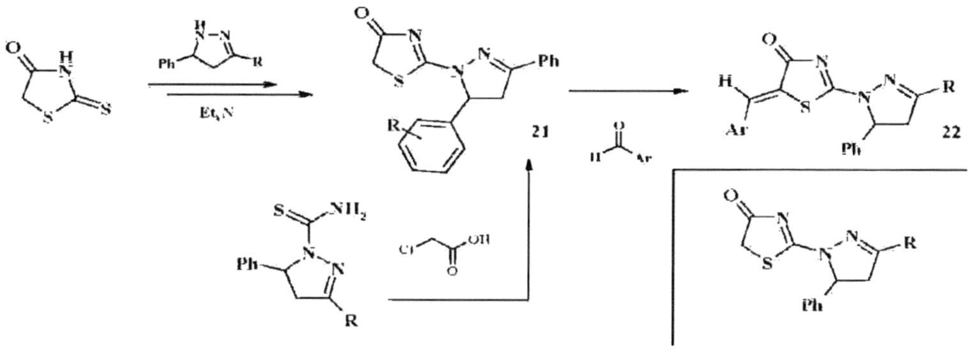

**Figure 13.**
*"Fixation" of hydrazone fragment for thiazolidinone-pyrazoline hybrid synthesis.*

The hit compound from thiazolidinone-pyrazoline hybrids **21** showed inhibitory activity on the in vitro growth of *T.b. rhodesiense* ($IC_{50}$ = 12 μg/mL) and *Leishmania donovani* ($IC_{50}$> 30 μg/mL) and a higher influence on *Plasmodium falciparum* ($IC_{50}$ > 5 μg/mL) with cytotoxicity level $CC_{50}$> 90 μg/mL.

## 3. Fused heterocyclic molecules based on the core 4-thiazolidinone

Thiopyranothiazoles that frequently are synthesized in *hetero*-Diels-Alder reaction starting from 5-ene-thiazolidinones are considered as their fused mimetics, without Michael acceptor properties, though with saved pharmacological profiles (**Figure 14**) [90, 96, 156, 157].

So, various thiopyranothiazoles serve as a fruitful source of drug-like molecules that, unlike their precursors 5-ylidene-4-thiazolidinones, cannot be claimed as PAINS [99]. This class of fused thiazolidinone derivatives is characterized by a number of different biological activities [158], the most studied being the antitumor activity [96, 157, 159, 160]. Recently, antiparasitic properties of these polycyclic compounds have been also reported.

A series of N-substituted thiopyrano[2,3-*d*]thiazoles showed excellent inhibitory activity of *T.b. brucei* (bloodstream form) at the concentration of 10 μg/mL in vitro. The most promising compounds were 3-[2-(4-fluoro(chloro)phenyl)-2-oxoethyl]-3,5a,6,11b-tetrahydro-2*H*,5*H*-chromeno[4′,3′:4,5]thiopyrano[2,3-*d*]thiazol-2-ones and N-(4-chloro(ethylcarboxy)phenyl)-2-(2-oxo-5a-methyl-(5aRS,11bSR)-3,5a,6,11b-tetrahydro-2*H*,5*H*-chromeno[4′,3′:4,5]thiopyrano[2,3-*d*]thiazol-3-yl)-acetamides **23** (**Figure 15**) that inhibited more than 95% of parasite growth in the above concentration and near quarter at the concentration of 1 μg/mL [132].

Development of novel synthetic protocols for the thiopyrano[2,3-*d*]thiazoles and their modifications led to the synthesis of new spiro thiopyrano[2,3-*d*]thiazoles. A hit compound rel-(6′R,7′R)-7′-(3,4-dimethoxyphenyl)-1-(4-chlorophenyl)-3′,7′-dihydro-2*H*,2′*H*,5*H*-spiro[pyrolidin-3,6′-thiopyrano[2,3-*d*]thiazol]-2,2′,5-trione **24** (**Figure 16**), inhibiting growth of *T.b. brucei* and *T.b. gambiense* with the $IC_{50}$ values of 0.26 μM and 0.42 μM, respectively, was identified [161].

Effective and feasible method of functionalized thiazolothiopyrane core synthesis has been the utilization of norbornene as a dienophile with 5-ylidene-isorhodanines as heterodienes in the *hetero*-Diels-Alder reaction. Obtained 9-aryl(heteryl)-3,7-dithia-5-azatetracyclo[9.2.1.0$^{2,10}$.0$^{4,8}$]tetradecen-4(8)-ones-6 and their N-arylidene substituted analogues **25** (**Figure 17**) showed moderate trypanocidal activity. The most active representatives possessed $IC_{50}$ within

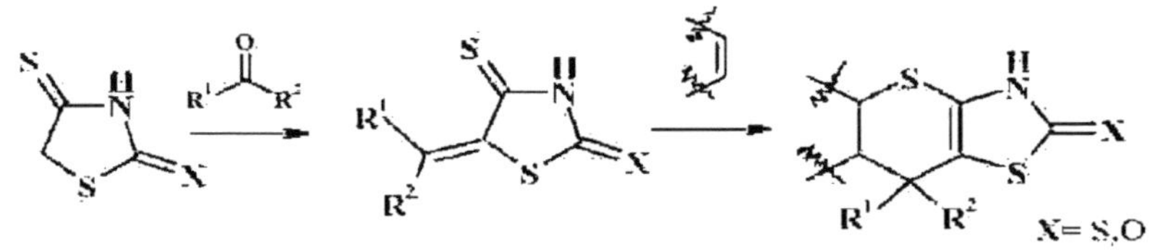


**Figure 14.**
*General scheme of thiopyranothiazole core formation.*

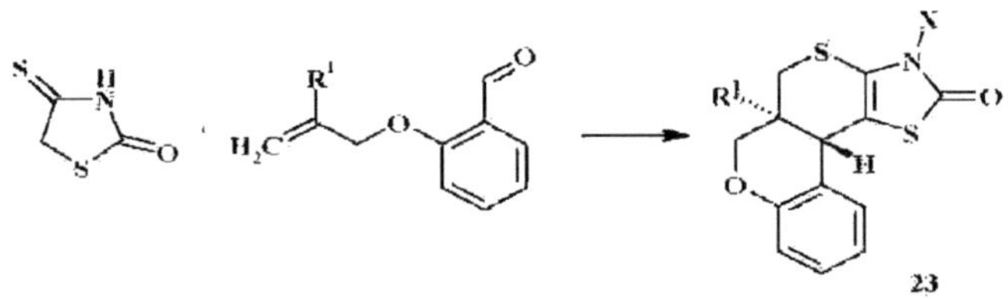

23

X = $CH_2$-CO-(4-F-$C_6H_4$);
$CH_2$-CO-(4-Cl-$C_6H_4$);
$CH_2$-CONH-(4-COOEt-$C_6H_4$);
$CH_2$-CONH-(4-Cl-$C_6H_4$)

**Figure 15.**
*Chromeno-thiopyrano-thiazolidinones as trypanocidal agents.*

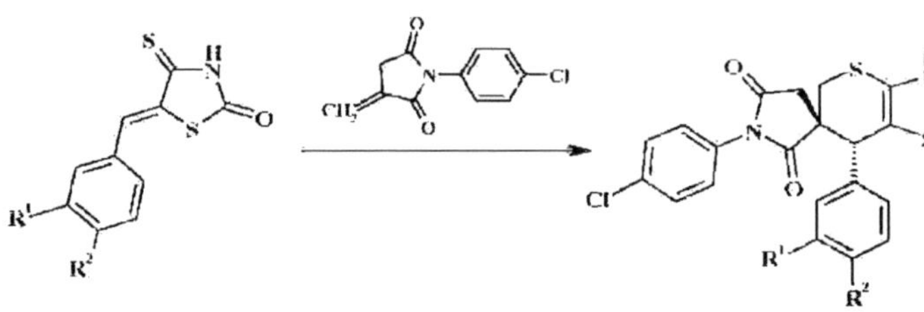

24 $R^1$ = $R^2$ = OMe
$IC_{50}$ = 0.26 μM (*T.b. brucei*),
0.42 μM (*T.b. gambiense*)

**Figure 16.**
*Synthesis of spiro thiopyrano[2,3-*d*]thiazole derivatives as trypanocidal agents.*

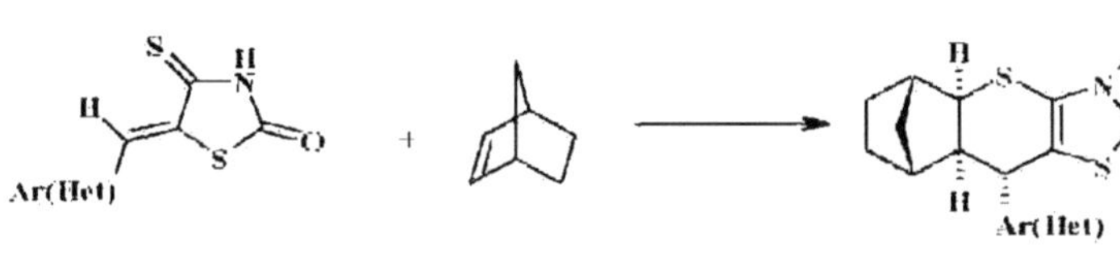

25

Ar = 4-Cl-$C_6H_4$;
Het = thiophen-2-yl;
X = H; $CH_2CONHC_6H_4$-4-OMe

**Figure 17.**
*Thiopyrano[2,3-*d*]thiazoles bearing norbornane moiety as antitrypanosomal agent.*

3.7–4.1 μM against *T.b. brucei*. Interesting was the dual antileukemic and trypanocidal effects observed for some thiopyranothiazoles bearing norbornane moiety that may be used for establishing the molecular mode of action for this class of compounds [118].

Comparable antitrypanosomal activity was observed for a series of isothiochromeno[4a,4-*d*][1,3]thiazoles **26** (**Figure 18**) in vitro against bloodstream forms of *T.b. brucei*. It should be mentioned that SAR analysis revealed the positive influence of N3-substituent for the trypanocidal activity. The same trend was found for the abovementioned tetracyclic thiopyrano[2,3-*d*]thiazoles **23** and thiopyranothiazoles with norbornane core **25**. Good trypanocidal properties along with a low acute toxicity in mice ($LD_{50}$: 240–480 mg/kg) for the isothiochromeno[4a,4-*d*][1,3] thiazole hits make such fused systems based on the thiazolidinone core attractive scaffolds for the discovery of antitrypanosomals [162].

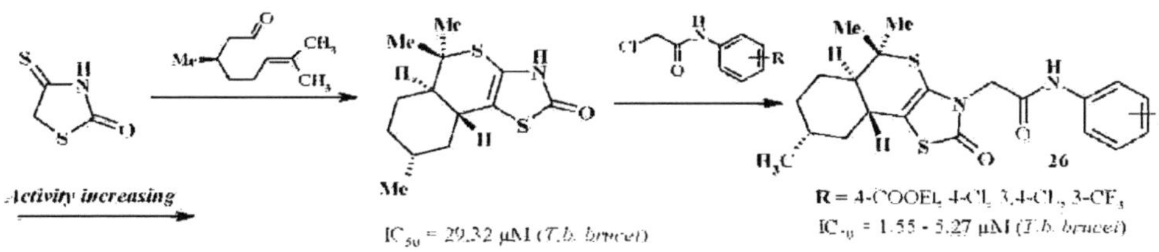


**Figure 18.**
*Isothiochromeno[4a,4-*d*][1,3]thiazoles as antitrypanosomal agents.*

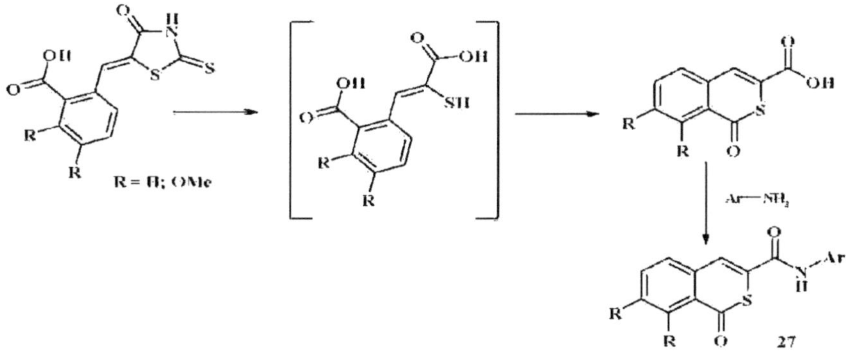


**Figure 19.**
*Rhodanine-based isothiocoumarine derivative synthesis.*

One more class of polycyclic fused molecules based on the thiazolidinone scaffold, being tested against *T.b. brucei*, were different 1-oxo-*1H*-2-benzothiopyran-3-carboxylic acids. The latter were synthesized in a result of heterocyclization of intermediates obtained in the hydrolysis reaction of 5-arylidenerhodanines with substituent in *ortho* position (**Figure 19**). Investigated amides did not exhibit significant antitrypanosomal effects except 1-oxo-1H-isothiochromene-3-carboxylic acid naphthalen-1-ylamide and 7,8-dimethoxy-1-oxo-1H-isothiochromene-3-carboxylic acid (4-sulfamoyl-phenyl)-amide **27** that inhibited growth of *T.b. brucei* bloodstream forms [119].

## 4. Conclusion

Thus, 4-thiazolidinone derivatives, especially thiazolidinone-bearing hybrids, as well as fused analogues are efficient compounds for the design of new antitrypanosomal agents within different drug design strategies. Thiazolidinone derivatives are more active than the known thiosemicarbazone analogues. Moreover, they can be used as starting compounds for the design and development of non-thiazolidinone compounds with trypanocidal activity. In addition, there are many active anticancer agents among 4-thiazolidinones with trypanocidal properties, and some active antitrypanosomal 4-thiazolidinones can be interesting for the search for new antimalarial and antileishmanial agents.

## Conflict of interest

The authors declare no conflict of interest.

**Author details**

Anna Kryshchyshyn[1], Danylo Kaminskyy[1], Philippe Grellier[2*] and Roman Lesyk[1,3*]

1 Department of Pharmaceutical, Organic and Bioorganic Chemistry, Danylo Halytsky Lviv National Medical University, Lviv, Ukraine

2 National Museum of Natural History, UMR 7245 CNRS-MNHN, Team PPL, CP 52, Paris, France

3 Faculty of Medicine, Department of Public Health, Dietetics and Lifestyle Disorders, University of Information Technology and Management in Rzeszow, Rzeszow, Poland

*Address all correspondence to: philippe.grellier@mnhn.fr
and dr_r_lesyk@org.lviv.net

## References

[1] Neglected Tropical Diseases [Internet]. Available from: https://www.who.int/neglected_diseases/diseases/en/ [Accessed: 03 February 2020]

[2] Hotez PJ, Fenwick A, Savioli L, Molyneux DH. Rescuing the bottom billion through control of neglected tropical diseases. The Lancet. 2009;**373**:1570-1575. DOI: 10.1016/S0140-6736(09)60233-6

[3] Filardy A, Guimarães-Pinto K, Nunes M, Zukeram K, Fliess L, Pereira L, et al. Human kinetoplastid protozoan infections: Where are we going next? Frontiers in Immunology. 2018;**9**:1493. DOI: 10.3389/fimmu.2018.01493

[4] Büscher P, Cecchi G, Jamonneau V, Priotto G. Human African trypanosomiasis. The Lancet. 2017;**390**:2397-2409. DOI: 10.1016/S0140-6736(17)31510-6

[5] Trypanosomiasis, Human African (Sleeping Sickness). Available from: https://www.who.int/news-room/fact-sheets/detail/trypanosomiasis-human-african-(sleeping-sickness) [Accessed: 03 February 2020]

[6] Njiokou F, Nimpaye H, Simo G, Njitchouang G, Asonganyi T, Cuny G, et al. Domestic animals as potential reservoir hosts of *Trypanosoma brucei* gambiense in sleeping sickness foci in Cameroon. Parasite. 2010;**17**(1):61-66. DOI: 10.1051/parasite/2010171061

[7] Simo G, Rayaisse J. Challenges facing the elimination of sleeping sickness in west and Central Africa: Sustainable control of animal trypanosomiasis as an indispensable approach to achieve the goal. Parasites & Vectors. 2015;**8**(1):640. DOI: 10.1186/s13071-015-1254-y

[8] Simarro P, Diarra A, Ruiz Postigo J, Franco J, Jannin J. The human African trypanosomiasis control and surveillance programme of the World Health Organization 2000-2009: The way forward. PLoS Neglected Tropical Diseases. 2011;**5**(2):e1007. DOI: 10.1371/journal.pntd.0001007

[9] Barrett M, Boykin D, Brun R, Tidwell R. Human African trypanosomiasis: Pharmacological re-engagement with a neglected disease. British Journal of Pharmacology. 2007;**152**(8):1155-1171. DOI: 10.1038/sj.bjp.0707354

[10] Barrett M, Croft S. Management of trypanosomiasis and leishmaniasis. British Medical Bulletin. 2012;**104**(1): 175-196. DOI: 10.1093/bmb/lds031

[11] Echeverria L, Morillo C. American trypanosomiasis (Chagas disease). Infectious Disease Clinics of North America. 2019;**33**(1):119-134. DOI: 10.1016/j.idc.2018.10.015

[12] Bern C, Montgomery S. An estimate of the burden of Chagas disease in the United States. Clinical Infectious Diseases. 2009;**49**(5):e52-e54. DOI: 10.1086/605091

[13] Chagas disease (American trypanosomiasis). Available from: https://www.who.int/en/news-room/fact-sheets/detail/chagas-disease-(american-trypanosomiasis) [Accessed: 03 February 2020]

[14] Tzizik D, Borchardt R. Chagas disease. Journal of the American Academy of Physician Assistants. 2018;**31**(12):30-33. DOI: 10.1097/01.jaa.0000547749.92933.6a

[15] Tarleton R. Parasite persistence in the aetiology of Chagas disease. International Journal for Parasitology. 2001;**31**(5-6): 550-554. DOI: 10.1016/s0020-7519(01)00158-8

[16] Pérez-Molina J, Molina I. Chagas disease. The Lancet.

2018;**391**(10115):82-94. DOI: 10.1016/s0140-6736(17)31612-4

[17] Rassi A, Marin-Neto J. Chagas disease. The Lancet. 2010;**375**(9723):1388-1402. DOI: 10.1016/s0140-6736(10)60061-x

[18] Nunes M, Dones W, Morillo C, Encina J, Ribeiro A. Chagas disease. Journal of the American College of Cardiology. 2013;**62**(9):767-776. DOI: 10.1016/j.jacc.2013.05.046

[19] Simarro P, Franco J, Diarra A, Postigo J, Jannin J. Update on field use of the available drugs for the chemotherapy of human African trypanosomiasis. Parasitology. 2012;**139**(7):842-846. DOI: 10.1017/s0031182012000169

[20] Kennedy P. Clinical features, diagnosis, and treatment of human African trypanosomiasis (sleeping sickness). The Lancet Neurology. 2013;**12**(2):186-194. DOI: 10.1016/s1474-4422(12)70296-x

[21] Franco J, Simarro P, Diarra A, Ruiz-Postigo JA, Samo M, Jannin J. Monitoring the use of nifurtimox-eflornithine combination therapy (NECT) in the treatment of second stage gambiense human African trypanosomiasis. Research and Reports in Tropical Medicine. 2012;**3**:93. DOI: 10.2147/rrtm.s34399

[22] Brun R, Blum J, Chappuis F, Burri C. Human African trypanosomiasis. The Lancet. 2010;**375**(9709):148-159. DOI: 10.1016/s0140-6736(09)60829-1

[23] Kennedy P. Human African trypanosomiasis of the CNS: Current issues and challenges. Journal of Clinical Investigation. 2004;**113**(4):496-504. DOI: 10.1172/jci200421052

[24] Priotto G, Pinoges L, Fursa I, Burke B, Nicolay N, Grillet G, et al. Safety and effectiveness of first line eflornithine for *Trypanosoma brucei* gambiense sleeping sickness in Sudan: Cohort study. BMJ. 2008;**336**(7646):705-708. DOI: 10.1136/bmj.39485.592674.be

[25] Chappuis F, Udayraj N, Stietenroth K, Meussen A, Bovier P. Eflornithine is safer than Melarsoprol for the treatment of second-stage *Trypanosoma brucei* gambiense human African trypanosomiasis. Clinical Infectious Diseases. 2005;**41**(5):748-751. DOI: 10.1086/432576

[26] Burri C, Nkunku S, Merolle A, Smith T, Blum J, Brun R. Efficacy of new, concise schedule for melarsoprol in treatment of sleeping sickness caused by *Trypanosoma brucei* gambiense: A randomised trial. The Lancet. 2000;**355**(9213):1419-1425. DOI: 10.1016/s0140-6736(00)02141-3

[27] Kuepfer I, Schmid C, Allan M, Edielu A, Haary E, Kakembo A, et al. Safety and efficacy of the 10-Day melarsoprol schedule for the treatment of second stage rhodesiense sleeping sickness. PLoS Neglected Tropical Diseases. 2012;**6**(8):e1695. DOI: 10.1371/journal.pntd.0001695

[28] Mesu V, Kalonji W, Bardonneau C, Mordt O, Blesson S, Simon F, et al. Oral fexinidazole for late-stage African *Trypanosoma brucei* gambiense trypanosomiasis: A pivotal multicentre, randomised, non-inferiority trial. The Lancet. 2018;**391**(10116):144-154. DOI: 10.1016/s0140-6736(17)32758-7

[29] European Medicines Agency Committee for Medicinal Products for Human Use. Assessment Report Fexinidazole Winthrop. 2018. Available from: https://www.ema.europa.eu/en/documents/medicine-outside-eu/fexinidazole-winthrop-assessment-report_en-0.pdf [Accessed: 03 February 2020]

[30] Pelfrene E, Harvey Allchurch M, Ntamabyaliro N, Nambasa V,

Ventura F, Nagercoil N, et al. The European medicines Agency's scientific opinion on oral fexinidazole for human African trypanosomiasis. PLoS Neglected Tropical Diseases. 2019;**13**(6):e0007381. DOI: 10.1371/journal.pntd.0007381

[31] Lindner A, Lejon V, Chappuis F, Seixas J, Kazumba L, Barrett M, et al. New WHO guidelines for treatment of gambiense human African trypanosomiasis including fexinidazole: Substantial changes for clinical practice. The Lancet Infectious Diseases. 2020;**20**(2):e38-e46. DOI: 10.1016/s1473-3099(19)30612-7

[32] Deeks E. Fexinidazole: First global approval. Drugs. 2019;**79**(2):215-220. DOI: 10.1007/s40265-019-1051-6

[33] Croft S, Barrett M, Urbina J. Chemotherapy of trypanosomiases and leishmaniasis. Trends in Parasitology. 2005;**21**(11):508-512. DOI: 10.1016/j.pt.2005.08.026

[34] Edwards D. Nitroimidazole drugs-action and resistance mechanisms I. Mechanism of action. Journal of Antimicrobial Chemotherapy. 1993;**31**(1):9-20. DOI: 10.1093/jac/31.1.9

[35] Viotti R, Vigliano C, Lococo B, Alvarez M, Petti M, Bertocchi G, et al. Side effects of benznidazole as treatment in chronic Chagas disease: Fears and realities. Expert Review of Anti-Infective Therapy. 2009;**7**(2):157-163. DOI: 10.1586/14787210.7.2.157

[36] Bern C. Chagas' disease. New England Journal of Medicine. 2015;**373**(5):456-466. DOI: 10.1056/nejmra1410150

[37] Apt W. Current and developing therapeutic agents in the treatment of Chagas disease. Drug Design, Development and Therapy. 2010;**4**:243-253. DOI: 10.2147/dddt.s8338

[38] Pierce R, MacDougall J, Leurs R, Costi M. The future of drug development for neglected tropical diseases: How the European Commission can continue to make a difference. Trends in Parasitology. 2017;**33**(8):581-583. DOI: 10.1016/j.pt.2017.04.007

[39] Ioset J-R, Chatelain E. Drug discovery and development for neglected diseases: The DNDi model. Drug Design, Development and Therapy. 2011;**5**:175-183. DOI: 10.2147/dddt.s16381

[40] Werbovetz K. Target-based drug discovery for malaria, Leishmaniasis, and trypanosomiasis. Current Medicinal Chemistry. 2000;**7**(8):835-860. DOI: 10.2174/0929867003374615

[41] Chatelain E, Ioset J. Phenotypic screening approaches for Chagas disease drug discovery. Expert Opinion on Drug Discovery. 2018;**13**(2):141-153. DOI: 10.1080/17460441.2018.1417380

[42] Kryshchyshyn A, Kaminskyy D, Grellier P, Lesyk R. Trends in research of antitrypanosomal agents among synthetic heterocycles. European Journal of Medicinal Chemistry. 2014;**85**:51-64. DOI: 10.1016/j.ejmech.2014.07.092

[43] Wermuth C, editor. The Practice of Medicinal Chemistry. 3rd ed. San Diego, CA, USA: Academic Press; 2011. p. 982

[44] Paliwal S. Neglected disease—African sleeping sickness: Recent synthetic and modeling advances. Scientia Pharmaceutica. 2011;**79**(3):389-428. DOI: 10.3797/scipharm.1012-08

[45] Nwaka S, Hudson A. Innovative lead discovery strategies for tropical diseases. Nature Reviews. Drug Discovery. 2006;**5**(11):941-955. DOI: 10.1038/nrd2144

[46] Gilbert I. Drug discovery for neglected diseases: Molecular

target-based and phenotypic approaches. Journal of Medicinal Chemistry. 2013;**56**(20):7719-7726. DOI: 10.1021/jm400362b

[47] Nwaka S, Ramirez B, Brun R, Maes L, Douglas F, Ridley R. Advancing drug innovation for neglected diseases—Criteria for lead progression. PLoS Neglected Tropical Diseases. 2009;**3**(8):e440. DOI: 10.1371/journal.pntd.0000440

[48] Magarinos M, Carmona S, Crowther G, Ralph S, Roos D, Shanmugam D, et al. TDR targets: A chemogenomics resource for neglected diseases. Nucleic Acids Research. 2011;**40**(D1):D1118-D1127. DOI: 10.1093/nar/gkr1053

[49] Beatriz Vermelho A, Giovanni De Simone S, Masini d'Avila-Levy C, Luis Souza do Santos A, Cristina Nogueira de Melo A, Paes Silva F, et al. Trypanosomatidae peptidases: A target for drugs development. Current Enzyme Inhibition. 2007;**3**(1):19-48. DOI: 10.2174/157340807779815468

[50] Duschak V, Couto A. Cruzipain, the major cysteine protease of *Trypanosoma cruzi*: A sulfated glycoprotein antigen as relevant candidate for vaccine development and drug target. A review. Current Medicinal Chemistry. 2009;**16**(24):3174-3202. DOI: 10.2174/092986709788802971

[51] Authié E, Boulangé A, Muteti D, Lalmanach G, Gauthier F, Musoke A. Immunisation of cattle with cysteine proteinases of *Trypanosoma congolense*: Targeting the disease rather than the parasite. International Journal for Parasitology. 2001;**31**(13):1429-1433. DOI: 10.1016/s0020-7519(01)00266-1

[52] Caler E. Oligopeptidase B-dependent signaling mediates host cell invasion by *Trypanosoma cruzi*. The EMBO Journal. 1998;**17**(17):4975-4986. DOI: 10.1093/emboj/17.17.4975

[53] Morty R, Lonsdale-Eccles J, Mentele R, Auerswald E, Coetzer T. Trypanosome-derived oligopeptidase B is released into the plasma of infected rodents, where it persists and retains full catalytic activity. Infection and Immunity. 2001;**69**(4):2757-2761. DOI: 10.1128/iai.69.4.2757-2761.2001

[54] Santana J, Grellier P, Schrével J, Teixeira A. A *Trypanosoma cruzi*-secreted 80 kDa proteinase with specificity for human collagen types I and IV. Biochemical Journal. 1997;**325**(1): 129-137. DOI: 10.1042/bj3250129

[55] Grellier P, Vendeville S, Joyeau R, Bastos I, Drobecq H, Frappier F, et al. *Trypanosoma cruzi* prolyl oligopeptidase Tc80 is involved in nonphagocytic mammalian cell invasion by Trypomastigotes. Journal of Biological Chemistry. 2001;**276**(50):47078-47086. DOI: 10.1074/jbc.m106017200

[56] Hall B, Bot C, Wilkinson S. Nifurtimox activation by Trypanosomal type I nitroreductases generates cytotoxic nitrile metabolites. Journal of Biological Chemistry. 2011;**286**(15):13088-13095. DOI: 10.1074/jbc.m111.230847

[57] Smith T, Young B, Denton H, Hughes D, Wagner G. First small molecular inhibitors of *T. brucei* dolicholphosphate mannose synthase (DPMS), a validated drug target in African sleeping sickness. Bioorganic and Medicinal Chemistry Letters. 2009;**19**(6):1749-1752. DOI: 10.1016/j.bmcl.2009.01.083

[58] Gangjee A, Jain H, Kurup S. Recent advances in classical and non-classical antifolates as antitumor and antiopportunistic infection agents: Part I. Anti-Cancer Agents in Medicinal Chemistry. 2007;**7**(5):524-542. DOI: 10.2174/187152007781668724

[59] Gilbert I. Inhibitors of dihydrofolate reductase in leishmania and trypanosomes. Biochimica et Biophysica Acta (BBA)—Molecular Basis of

Disease. 2002;**1587**(2-3):249-257. DOI: 10.1016/s0925-4439(02)00088-1

[60] Krieger S, Schwarz W, Ariyanayagam M, Fairlamb A, Krauth-Siegel R, Clayton C. Trypanosomes lacking trypanothione reductase are avirulent and show increased sensitivity to oxidative stress. Molecular Microbiology. 2002;**35**(3):542-552. DOI: 10.1046/j.1365-2958.2000.01721.x

[61] Spinks D, Shanks E, Cleghorn L, McElroy S, Jones D, James D, et al. Investigation of Trypanothione reductase as a drug target in *Trypanosoma brucei*. ChemMedChem. 2009;**4**(12):2060-2069. DOI: 10.1002/cmdc.200900262

[62] Parsons M, Worthey E, Ward P, Mottram J. Journal search results—Cite this for me. BMC Genomics. 2005;**6**(1):127. DOI: 10.1186/1471-2164-6-127

[63] Naula C, Parsons M, Mottram J. Protein kinases as drug targets in trypanosomes and Leishmania. Biochimica et Biophysica Acta (BBA)—Proteins and Proteomics. 2005;**1754**(1-2):151-159. DOI: 10.1016/j.bbapap.2005.08.018

[64] Olivares-Illana V, Pérez-Montfort R, López-Calahorra F, Costas M, Rodríguez-Romero A, Tuena de Gómez-Puyou M, et al. Structural differences in triosephosphate isomerase from different species and discovery of a multitrypanosomatid inhibitor. Biochemistry. 2006;**45**(8):2556-2560. DOI: 10.1021/bi0522293

[65] Verlinde C, Hannaert V, Blonski C, Willson M, Périé J, Fothergill-Gilmore L, et al. Glycolysis as a target for the design of new anti-trypanosome drugs. Drug Resistance Updates. 2001;**4**(1):50-65. DOI: 10.1054/drup.2000.0177

[66] Ostoa-Saloma P, Garza-Ramos G, Ramirez J, Becker I, Berzunza M, Landa A, et al. Cloning, expression, purification and characterization of triosephosphate isomerase from *Trypanosoma cruzi*. European Journal of Biochemistry. 1997;**244**(3):700-705. DOI: 10.1111/j.1432-1033.1997.00700.x

[67] Gabelli S, McLellan J, Montalvetti A, Oldfield E, Docampo R, Amzel L. Structure and mechanism of the farnesyl diphosphate synthase from *Trypanosoma cruzi*: Implications for drug design. Proteins: Structure, Function, and Bioinformatics. 2005;**62**(1):80-88. DOI: 10.1002/prot.20754

[68] Montalvetti A, Bailey B, Martin M, Severin G, Oldfield E, Docampo R. Bisphosphonates are potent inhibitors of *Trypanosoma cruzi* farnesyl pyrophosphate synthase. Journal of Biological Chemistry. 2001;**276**(36):33930-33937. DOI: 10.1074/jbc.m103950200

[69] Aripirala S, Szajnman S, Jakoncic J, Rodriguez J, Docampo R, Gabelli S, et al. Design, synthesis, calorimetry, and crystallographic analysis of 2-alkylaminoethyl-1,1-bisphosphonates as inhibitors of *Trypanosoma cruzi* farnesyl diphosphate synthase. Journal of Medicinal Chemistry. 2012;**55**(14):6445-6454. DOI: 10.1021/jm300425y

[70] de Koning H, Gould M, Sterk G, Tenor H, Kunz S, Luginbuehl E, et al. Pharmacological validation of *Trypanosoma brucei* phosphodiesterases as novel drug targets. The Journal of Infectious Diseases. 2012;**206**(2):229-237. DOI: 10.1093/infdis/jir857

[71] Kunz S, Beavo J, D'Angelo M, Flawia M, Francis S, Johner A, et al. Cyclic nucleotide specific phosphodiesterases of the kinetoplastida: A unified nomenclature. Molecular and Biochemical Parasitology. 2006;**145**(1):133-135. DOI: 10.1016/j.molbiopara.2005.09.018

[72] Patel G, Karver C, Behera R, Guyett P, Sullenberger C, Edwards P, et al. Kinase scaffold repurposing for neglected disease drug discovery: Discovery of an efficacious, lapatinib-derived lead compound for trypanosomiasis. Journal of Medicinal Chemistry. 2013;**56**(10):3820-3832. DOI: 10.1021/jm400349k

[73] Ochiana S, Pandarinath V, Wang Z, Kapoor R, Ondrechen M, Ruben L, et al. The human aurora kinase inhibitor danusertib is a lead compound for anti-trypanosomal drug discovery via target repurposing. European Journal of Medicinal Chemistry. 2013;**62**:777-784. DOI: 10.1016/j.ejmech.2012.07.038

[74] Katiyar S, Kufareva I, Behera R, Thomas S, Ogata Y, Pollastri M, et al. Lapatinib-binding protein kinases in the African trypanosome: Identification of cellular targets for kinase-directed chemical scaffolds. PLoS One. 2013;**8**(2):e56150. DOI: 10.1371/journal.pone.0056150

[75] Bailey S, Smith K, Fairlamb A, Hunter W. Substrate interactions between trypanothione reductase and N1-glutathionylspermidine disulphide at 0.28-nm resolution. European Journal of Biochemistry. 1993;**213**(1):67-75. DOI: 10.1111/j.1432-1033.1993.tb17734.x

[76] Pollastri M, Campbell R. Target repurposing for neglected diseases. Future Medicinal Chemistry. 2011;**3**(10):1307-1315. DOI: 10.4155/fmc.11.92

[77] Lepesheva G, Waterman M. CYP51—The omnipotent P450. Molecular and Cellular Endocrinology. 2004;**215**(1-2):165-170. DOI: 10.1016/j.mce.2003.11.016

[78] Lepesheva G, Waterman M. Sterol 14α-demethylase cytochrome P450 (CYP51), a P450 in all biological kingdoms. Biochimica et Biophysica Acta (BBA)—General Subjects. 2007;**1770**(3):467-477. DOI: 10.1016/j.bbagen.2006.07.018

[79] Lepesheva G, Friggeri L, Waterman M. CYP51 as drug targets for fungi and protozoan parasites: Past, present and future. Parasitology. 2018;**145**(14):1820-1836. DOI: 10.1017/s0031182018000562

[80] Haasen D, Schopfer U, Antczak C, Guy C, Fuchs F, Selzer P. How phenotypic screening influenced drug discovery: Lessons from five years of practice. ASSAY and Drug Development Technologies. 2017;**15**(6):239-246. DOI: 10.1089/adt.2017.796

[81] Moffat J, Vincent F, Lee J, Eder J, Prunotto M. Opportunities and challenges in phenotypic drug discovery: An industry perspective. Nature Reviews. Drug Discovery. 2017;**16**(8):531-543. DOI: 10.1038/nrd.2017.111

[82] Swinney D. Phenotypic vs. target-based drug discovery for first-in-class medicines. Clinical Pharmacology and Therapeutics. 2013;**93**(4):299-301. DOI: 10.1038/clpt.2012.236

[83] Peña I, Pilar Manzano M, Cantizani J, Kessler A, Alonso-Padilla J, Bardera A, et al. New compound sets identified from high throughput phenotypic screening against three kinetoplastid parasites: An open resource. Scientific Reports. 2015;**5**(1):8771. DOI: 10.1038/srep08771

[84] Sykes M, Avery V. Approaches to protozoan drug discovery: Phenotypic screening. Journal of Medicinal Chemistry. 2013;**56**(20):7727-7740. DOI: 10.1021/jm4004279

[85] Donelson J, Gardner M, El-Sayed N. More surprises from kinetoplastida. Proceedings of the National Academy of Sciences. 1999;**96**(6):2579-2581. DOI: 10.1073/pnas.96.6.2579

[86] Stuart K, Brun R, Croft S, Fairlamb A, Gürtler R, McKerrow J, et al. Kinetoplastids: Related protozoan pathogens, different diseases. Journal of Clinical Investigation. 2008;**118**(4):1301-1310. DOI: 10.1172/jci33945

[87] Medina-Franco J, Giulianotti M, Welmaker G, Houghten R. Shifting from the single to the multitarget paradigm in drug discovery. Drug Discovery Today. 2013;**18**(9-10):495-501. DOI: 10.1016/j.drudis.2013.01.008

[88] Bansal Y, Silakari O. Multifunctional compounds: Smart molecules for multifactorial diseases. European Journal of Medicinal Chemistry. 2014;**76**:31-42. DOI: 10.1016/j.ejmech.2014.01.060

[89] Tripathi A, Gupta S, Fatima G, Sonar P, Verma A, Saraf S. 4-Thiazolidinones: The advances continue.... European Journal of Medicinal Chemistry. 2014;**72**:52-77. DOI: 10.1016/j.ejmech.2013.11.017

[90] Lesyk R, Zimenkovsky B. 4-thiazolidones: Centenarian history, current status and perspectives for modern organic and medicinal chemistry. Current Organic Chemistry. 2004;**8**(16):1547-1577. DOI: 10.2174/1385272043369773

[91] Tomasic T, Masic L. Rhodanine as a privileged scaffold in drug discovery. Current Medicinal Chemistry. 2009;**16**(13):1596-1629. DOI: 10.2174/092986709788186200

[92] Lesyk R, Zimenkovsky B, Kaminskyy D, Kryshchyshyn A, Havryluk D, Atamanyuk D, et al. Thiazolidinone motif in anticancer drug discovery. Experience of DH LNMU medicinal chemistry scientific group. Biopolymers and Cell. 2011;**27**(2):107-117. DOI: 10.7124/bc.000089

[93] Verma A, Saraf S. 4-thiazolidinone—A biologically active scaffold. European Journal of Medicinal Chemistry. 2008;**43**(5):897-905. DOI: 10.1016/j.ejmech.2007.07.017

[94] Welsch M, Snyder S, Stockwell B. Privileged scaffolds for library design and drug discovery. Current Opinion in Chemical Biology. 2010;**14**(3):347-361. DOI: 10.1016/j.cbpa.2010.02.018

[95] Mendgen T, Steuer C, Klein C. Privileged scaffolds or promiscuous binders: A comparative study on Rhodanines and related heterocycles in medicinal chemistry. Journal of Medicinal Chemistry. 2012;**55**(2):743-753. DOI: 10.1021/jm201243p

[96] Kaminskyy D, Kryshchyshyn A, Lesyk R. 5-Ene-4-thiazolidinones—An efficient tool in medicinal chemistry. European Journal of Medicinal Chemistry. 2017;**140**:542-594. DOI: 10.1016/j.ejmech.2017.09.031

[97] Kaminskyy D, Kryshchyshyn A, Lesyk R. Recent developments with rhodanine as a scaffold for drug discovery. Expert Opinion on Drug Discovery. 2017;**12**(12):1233-1252. DOI: 10.1080/17460441.2017.1388370

[98] Oh S, Park S. A design strategy for drug-like polyheterocycles with privileged substructures for discovery of specific small-molecule modulators. Chemical Communications. 2011;**47**(48):12754-12761. DOI: 10.1039/c1cc14042f

[99] Baell J, Holloway G. New substructure filters for removal of pan assay interference compounds (PAINS) from screening libraries and for their exclusion in bioassays. Journal of Medicinal Chemistry. 2010;**53**(7):2719-2740. DOI: 10.1021/jm901137j

[100] Baell J, Walters M. Chemistry: Chemical con artists foil drug discovery. Nature. 2014;**513**(7519):481-483. DOI: 10.1038/513481a

[101] Aldrich C, Bertozzi C, Georg G, Kiessling L, Lindsley C, Liotta D, et al. The ecstasy and agony of assay

interference compounds. ACS Central Science. 2017;**3**(3):143-147. DOI: 10.1021/acscentsci.7b00069

[102] Parvez S, Long M, Poganik J, Aye Y. Redox signaling by reactive electrophiles and oxidants. Chemical Reviews. 2018;**118**(18):8798-8888. DOI: 10.1021/acs.chemrev.7b00698

[103] Pinson J, Schmidt-Kittler O, Frazzetto M, Zheng Z, Jennings I, Kinzler K, et al. Synthesis and pharmacological evaluation of 4-iminothiazolidinones for inhibition of PI3 kinase. Australian Journal of Chemistry. 2012;**65**(10):1396-1404. DOI: 10.1071/ch12140

[104] Smelcerovic Z, Veljkovic A, Kocic G, Yancheva D, Petronijevic Z, Anderluh M, et al. Xanthine oxidase inhibitory properties and anti-inflammatory activity of 2-amino-5-alkylidene-thiazol-4-ones. Chemico-Biological Interactions. 2015;**229**:73-81. DOI: 10.1016/j.cbi.2015.01.022

[105] Brown F. 4-Thiazolidinones. Chemical Reviews. 1961;**61**(5):463-521. DOI: 10.1021/cr60213a002

[106] Singh S, Parmar S, Raman K, Stenberg V. Chemistry and biological activity of thiazolidinones. Chemical Reviews. 1981;**81**(2):175-203. DOI: 10.1021/cr00042a003

[107] Newkome G, Nayak A. 4-Thiazolidinones. Advances in Heterocyclic Chemistry. 1980;**25**:83-112. DOI: 10.1016/S0065-2725(08)60690-X

[108] Calogeropoulou T, Magoulas G, Pöhner I, Panecka-Hofman J, Ferrari S, Costi M, et al. Hits and lead discovery in the identification of new drugs against the trypanosomatidic infections. In: Medicinal Chemistry of Neglected and Tropical Diseases: Advances in the Design and Synthesis of Antimicrobial Agents. Florida, USA: CRC Press Taylor & Francis Group; 2019. pp. 185-231

[109] Du X, Guo C, Hansell E, Doyle P, Caffrey C, Holler T, et al. Synthesis and structure–activity relationship study of potent trypanocidal thio semicarbazone inhibitors of the trypanosomal cysteine protease cruzain. Journal of Medicinal Chemistry. 2002;**45**(13):2695-2707. DOI: 10.1021/jm010459j

[110] Fatondji H, Kpoviessi S, Gbaguidi F, Bero J, Hannaert V, Quetin-Leclercq J, et al. Structure–activity relationship study of thiosemicarbazones on an African trypanosome: *Trypanosoma brucei brucei*. Medicinal Chemistry Research. 2012;**22**(5):2151-2162. DOI: 10.1007/s00044-012-0208-6

[111] Pérez-Rebolledo A, Teixeira L, Batista A, Mangrich A, Aguirre G, Cerecetto H, et al. 4-Nitroacetophenone-derived thiosemicarbazones and their copper(II) complexes with significant in vitro anti-trypanosomal activity. European Journal of Medicinal Chemistry. 2008;**43**(5):939-948. DOI: 10.1016/j.ejmech.2007.06.020

[112] Parrilha G, Dias R, Rocha W, Mendes I, Benítez D, Varela J, et al. 2-Acetylpyridine-and2-benzoylpyridine-derived thiosemicarbazones and their antimony(III) complexes exhibit high anti-trypanosomal activity. Polyhedron. 2012;**31**(1):614-621. DOI: 10.1016/j.poly.2011.10.018

[113] Mallari J, Shelat A, Kosinski A, Caffrey C, Connelly M, Zhu F, et al. Discovery of trypanocidal thiosemicarbazone inhibitors of rhodesain and TbcatB. Bioorganic and Medicinal Chemistry Letters. 2008;**18**(9):2883-2885. DOI: 10.1016/j.bmcl.2008.03.083

[114] Greenbaum D, Mackey Z, Hansell E, Doyle P, Gut J, Caffrey C, et al. Synthesis and structure–activity relationships of parasiticidal thiosemicarbazone cysteine protease inhibitors against *Plasmodium falciparum*, *Trypanosoma brucei*, and

*Trypanosoma cruzi*. Journal of Medicinal Chemistry. 2004;**47**(12):3212-3219. DOI: 10.1021/jm030549j

[115] Silva-Júnior E, Silva E, França P, Silva J, Barreto E, Silva E, et al. Design, synthesis, molecular docking and biological evaluation of thiophen-2-iminothiazolidine derivatives for use against *Trypanosoma cruzi*. Bioorganic & Medicinal Chemistry. 2016;**24**(18):4228-4240. DOI: 10.1016/j.bmc.2016.07.013

[116] Bhongade B, Talath S, Gadad R, Gadad A. Biological activities of imidazo[2,1-b][1,3,4]thiadiazole derivatives: A review. Journal of Saudi Chemical Society. 2016;**20**:S463-S475. DOI: 10.1016/j.jscs.2013.01.010

[117] Kryshchyshyn A, Devinyak O, Kaminskyy D, Grellier P, Lesyk R. Development of predictive QSAR models of 4-thiazolidinones antitrypanosomal activity using modern machine learning algorithms. Molecular Informatics. 2017;**37**(5):1700078. DOI: 10.1002/minf.201700078

[118] Kryshchyshyn A, Atamanyuk D, Kaminskyy D, Grellier P, Lesyk R. Investigation of anticancer and anti-parasitic activity of thiopyrano[2,3-d] thiazoles bearing norbornane moiety. Biopolymers and Cell. 2017;**33**(3):183-205. DOI: 10.7124/bc.00094f

[119] Kaminskyy D, Kryshchyshyn A, Nektegayev I, Vasylenko O, Grellier P, Lesyk R. Isothiocoumarin-3-carboxylic acid derivatives: Synthesis, anticancer and antitrypanosomal activity evaluation. European Journal of Medicinal Chemistry. 2014;**75**:57-66. DOI: 10.1016/j.ejmech.2014.01.028

[120] Leite A, Espíndola J, de Oliveira CM, de Oliveira Filho G. Privileged structures in the design of potential drug candidates for neglected diseases. Current Medicinal Chemistry. 2019;**26**(23):4323-4354. DOI: 10.2174/0929867324666171023163752

[121] Havrylyuk D, Zimenkovsky B, Karpenko O, Grellier P, Lesyk R. Synthesis of pyrazoline–thiazolidinone hybrids with trypanocidal activity. European Journal of Medicinal Chemistry. 2014;**85**:245-254. DOI: 10.1016/j.ejmech.2014.07.103

[122] Meunier B. Hybrid molecules with a dual mode of action: Dream or reality? Accounts of Chemical Research. 2008;**41**(1):69-77. DOI: 10.1021/ar7000843

[123] Viegas-Junior C, Barreiro EJ, Fraga CAM. Molecular hybridization: A useful tool in the design of new drug prototypes. Current Medicinal Chemistry. 2007;**14**(17):1829-1852. DOI: 10.2174/092986707781058805

[124] Ivasiv V, Albertini C, Gonçalves A, Rossi M, Bolognesi M. Molecular hybridization as a tool for designing multitarget drug candidates for complex diseases. Current Topics in Medicinal Chemistry. 2019;**19**(19):1694-1711. DOI: 10.2174/1568026619666190619115735

[125] Havrylyuk D, Zimenkovsky B, Vasylenko O, Day C, Smee D, Grellier P, et al. Synthesis and biological activity evaluation of 5-pyrazoline substituted 4-thiazolidinones. European Journal of Medicinal Chemistry. 2013;**66**:228-237. DOI: 10.1016/j.ejmech.2013.05.044

[126] Havrylyuk D, Zimenkovsky B, Zelisko N, Grellier P, Lesyk R. Antiprotozoal activity of non-condensed systems with thiazolidine and pyrazoline fragments. Clinical Pharmacy, Pharmacotherapy and Medicinal Standardization. 2010;**3-4**:62-67 (in Ukrainian)

[127] Havrylyuk D, Roman O, Lesyk R. Synthetic approaches, structure activity relationship and biological applications for pharmacologically attractive pyrazole/pyrazoline–thiazolidine-based hybrids. European Journal of Medicinal Chemistry. 2016;**113**:145-166. DOI: 10.1016/j.ejmech.2016.02.030

[128] Senkiv J, Finiuk N, Kaminskyy D, Havrylyuk D, Wojtyra M, Kril I, et al. 5-Ene-4-thiazolidinones induce apoptosis in mammalian leukemia cells. European Journal of Medicinal Chemistry. 2016;**117**:33-46. DOI: 10.1016/j.ejmech.2016.03.089

[129] Kaminskyy D, den Hartog G, Wojtyra M, Lelyukh M, Gzella A, Bast A, et al. Antifibrotic and anticancer action of 5-ene amino/iminothiazolidinones. European Journal of Medicinal Chemistry. 2016;**112**:180-195. DOI: 10.1016/j.ejmech.2016.02.011

[130] Kaminskyy D, Subtel'na I, Zimenkovsky B, Karpenko O, Gzella A, Lesyk R. Synthesis and evaluation of anticancer activity of 5-ylidene-4-aminothiazol-2(5H)-one derivatives. Medicinal Chemistry. 2015;**11**(6):517-530. DOI: 10.2174/1573406411666150211112049

[131] Zhang X, Li X, Li D, Qu G, Wang J, Loiseau P, et al. Ionic liquid mediated and promoted eco-friendly preparation of thiazolidinone and pyrimidine nucleoside–thiazolidinone hybrids and their antiparasitic activities. Bioorganic and Medicinal Chemistry Letters. 2009;**19**(22):6280-6283. DOI: 10.1016/j.bmcl.2009.09.101

[132] Kryshchyshyn A, Kaminskyy D, Zelisko N, Khyluk D, Grellier P, Lesyk R. The study of the antitrypanosomal activity of thiazolidinones and related heterocyclic systems. Journal of Organic and Pharmaceutical Chemistry. 2013;**11**(2):57-62 (in Ukrainian)

[133] Kaminskyy D, Khyluk D, Vasylenko O, Lesyk R. An efficient method for the transformation of 5-ylidenerhodanines into 2,3,5-trisubstituted-4-thiazolidinones. Tetrahedron Letters. 2012;**53**(5):557-559. DOI: 10.1016/j.tetlet.2011.11.095

[134] Kaminskyy D, Khyluk D, Vasylenko O, Zaprutko L, Lesyk R. A facile synthesis and anticancer activity evaluation of spiro [thiazolidinone-isatin] conjugates. Scientia Pharmaceutica. 2011;**79**(4):763-778. DOI: 10.3797/scipharm.1109-14

[135] Brem J, van Berkel S, Aik W, Rydzik A, Avison M, Pettinati I, et al. Rhodanine hydrolysis leads to potent thioenolate mediated metallo-β-lactamase inhibition. Nature Chemistry. 2014;**6**(12):1084-1090. DOI: 10.1038/nchem.2110

[136] Leite F, Santiago P, Froes T, da Silva FJ, da Silva S, Ximenes R, et al. Structure-guided discovery of thiazolidine-2,4-dione derivatives as a novel class of Leishmania major pteridine reductase 1 inhibitors. European Journal of Medicinal Chemistry. 2016;**123**:639-648. DOI: 10.1016/j.ejmech.2016.07.060

[137] Kaminskyy D, Zimenkovsky B, Lesyk R. Synthesis and in vitro anticancer activity of 2,4-azolidinedione-acetic acids derivatives. European Journal of Medicinal Chemistry. 2009;**44**(9):3627-3636. DOI: 10.1016/j.ejmech.2009.02.023

[138] Kaminskyy D, Lesyk R. Structure–anticancer activity relationships among 4-azolidinone-3-carboxylic acids derivatives. Biopolymers and Cell. 2010;**26**(2):136-145. DOI: 10.7124/bc.000150 (in Ukrainian)

[139] Holota S, Kryshchyshyn A, Derkach H, Trufin Y, Demchuk I, Gzella A, et al. Synthesis of 5-enamine-4-thiazolidinone derivatives with trypanocidal and anticancer activity. Bioorganic Chemistry. 2019;**86**:126-136. DOI: 10.1016/j.bioorg.2019.01.045

[140] Woodland A, Grimaldi R, Luksch T, Cleghorn L, Ojo K, Van Voorhis W, et al.

From on-target to off-target activity: Identification and optimisation of *Trypanosoma brucei* GSK3 inhibitors and their characterisation as anti-*Trypanosoma brucei* drug discovery lead molecules. ChemMedChem. 2013;**8**(7):1127-1137. DOI: 10.1002/cmdc.201300072

[141] de Oliveira FG, Cardoso M, Espíndola J, Oliveira e Silva D, Ferreira R, Coelho P, et al. Structural design, synthesis and pharmacological evaluation of thiazoles against *Trypanosoma cruzi*. European Journal of Medicinal Chemistry. 2017;**141**:346-361. DOI: 10.1016/j.ejmech.2017.09.047

[142] Pizzo C, Saiz C, Talevi A, Gavernet L, Palestro P, Bellera C, et al. Synthesis of 2-hydrazolyl-4-thiazolidinones based on multicomponent reactions and biological evaluation against *Trypanosoma cruzi*. Chemical Biology and Drug Design. 2011;**77**(3):166-172. DOI: 10.1111/j.1747-0285.2010.01071.x

[143] de Moraes GP, de Oliveira BM, Farias Santiago E, de Oliveira CM, Capistrano Costa N, Hernandes M, et al. New 1,3-thiazole derivatives and their biological and ultrastructural effects on *Trypanosoma cruzi*. European Journal of Medicinal Chemistry. 2016;**121**:387-398. DOI: 10.1016/j.ejmech.2016.05.050

[144] Scarim C, Jornada D, Chelucci R, de Almeida L, dos Santos J, Chung M. Current advances in drug discovery for Chagas disease. European Journal of Medicinal Chemistry. 2018;**155**:824-838. DOI: 10.1016/j.ejmech.2018.06.040

[145] Rodríguez G, Nargoli J, López A, Moyna G, Álvarez G, Fernández M, et al. Synthesis and in vivo proof of concept of a BODIPY-based fluorescent probe as a tracer for biodistribution studies of a new anti-Chagas agent. RSC Advances. 2017;**7**(13):7983-7989. DOI: 10.1039/c6ra27851e

[146] Álvarez G, Varela J, Cruces E, Fernández M, Gabay M, Leal S, et al. Identification of a new amide-containing thiazole as a drug candidate for treatment of Chagas' disease. Antimicrobial Agents and Chemotherapy. 2014;**59**(3):1398-1404. DOI: 10.1128/aac.03814-14

[147] Moreira D, Lima Leite A, Cardoso M, Srivastava R, Hernandes M, Rabello M, et al. Structural design, synthesis and structure-activity relationships of thiazolidinones with enhanced anti-*Trypanosoma cruzi* activity. ChemMedChem. 2013;**9**(1):177-188. DOI: 10.1002/cmdc.201300354

[148] Leite AC, Moreira DR, Cardoso MV, Hernandes MZ, Alves Pereira VR, Silva R, et al. Synthesis, cruzain docking, and in vitro studies of aryl-4-oxothiazolylhydrazones against *Trypanosoma cruzi*. ChemMedChem. 2007;**2**(9):1339-1345. DOI: 10.1002/cmdc.200700022

[149] Hernandes M, Rabello M, Leite A, Cardoso M, Moreira D, Brondani D, et al. Studies toward the structural optimization of novel thiazolylhydrazone-based potent antitrypanosomal agents. Bioorganic & Medicinal Chemistry. 2010;**18**(22): 7826-7835. DOI: 10.1016/j.bmc.2010.09.056

[150] Moreira D, Costa S, Hernandes M, Rabello M, de Oliveira FG, de Melo C, et al. Structural investigation of anti-*Trypanosoma cruzi* 2-Iminothiazolidin-4-ones allows the identification of agents with efficacy in infected mice. Journal of Medicinal Chemistry. 2012;**55**(24):10918-10936. DOI: 10.1021/jm301518v

[151] Leite A, de Lima R, Moreira D, Cardoso M, Gouveia de Brito A, Farias dos Santos L, et al. Synthesis, docking, and in vitro activity of

thiosemicarbazones, aminoacyl-thiosemicarbazides and acyl-thiazolidones against *Trypanosoma cruzi*. Bioorganic & Medicinal Chemistry. 2006;**14**(11):3749-3757. DOI: 10.1016/j.bmc.2006.01.034

[152] Seebacher W, Brun R, Weis R. New 4-aminobicyclo[2.2.2]octane derivatives and their activities against plasmodium falciparum and Trypanosoma b. rhodesiense. European Journal of Pharmaceutical Sciences. 2004;**21**(2-3):225-233. DOI: 10.1016/j.ejps.2003.10.011

[153] da Silva E, Oliveira e Silva D, Oliveira A, da Silva Mendes C, dos Santos T, da Silva A, et al. Design and synthesis of potent anti-*Trypanosoma cruzi* agents new thiazoles derivatives which induce apoptotic parasite death. European Journal of Medicinal Chemistry. 2017;**130**:39-50. DOI: 10.1016/j.ejmech.2017.02.026

[154] Caputto M, Ciccarelli A, Frank F, Moglioni A, Moltrasio G, Vega D, et al. Synthesis and biological evaluation of some novel 1-indanone thiazolylhydrazone derivatives as anti-*Trypanosoma cruzi* agents. European Journal of Medicinal Chemistry. 2012;**55**:155-163. DOI: 10.1016/j.ejmech.2012.07.013

[155] Kryshchyshyn A, Kaminskyy D, Karpenko O, Gzella A, Grellier P, Lesyk R. Thiazolidinone/thiazole based hybrids—New class of antitrypanosomal agents. European Journal of Medicinal Chemistry. 2019;**174**:292-308. DOI: 10.1016/j.ejmech.2019.04.052

[156] Lesyk R, Kaminskyy D, Vasylenko O, Atamanyuk D, Gzella A. Isorhodanine and thiorhodanine motifs in the synthesis of fused thiopyrano[2,3-d][1,3]thiazoles. Synlett. 2011;**10**:1385-1388. DOI: 10.1055/s-0030-1260765

[157] Kryshchyshyn A, Atamanyuk D, Lesyk R. Fused thiopyrano[2,3-d]thiazole derivatives as potential anticancer agents. Scientia Pharmaceutica. 2012;**80**(3):509-530. DOI: 10.3797/scipharm.1204-02

[158] Kryshchyshyn A, Roman O, Lozynskyi A, Lesyk R. Thiopyrano[2,3-d]thiazoles as new efficient scaffolds in medicinal chemistry. Scientia Pharmaceutica. 2018;**86**(2):26. DOI: 10.3390/scipharm86020026

[159] Atamanyuk D, Zimenkovsky B, Atamanyuk V, Nektegayev I, Lesyk R. Synthesis and biological activity of new thiopyrano[2,3-d]thiazoles containing a naphthoquinone moiety. Scientia Pharmaceutica. 2013;**81**(2):423-436. DOI: 10.3797/scipharm.1301-13

[160] Atamanyuk D, Zimenkovsky B, Atamanyuk V, Lesyk R. 5-Ethoxymethylidene-4-thioxo-2-thiazolidinone as versatile building block for novel biorelevant small molecules with thiopyrano[2,3-d][1,3]thiazole core. Synthetic Communications. 2013;**44**(2):237-244. DOI: 10.1080/00397911.2013.800552

[161] Zelisko N, Atamanyuk D, Vasylenko O, Grellier P, Lesyk R. Synthesis and antitrypanosomal activity of new 6,6,7-trisubstituted thiopyrano[2,3-d][1,3]thiazoles. Bioorganic and Medicinal Chemistry Letters. 2012;**22**(23):7071-7074. DOI: 10.1016/j.bmcl.2012.09.091

[162] Kryshchyshyn A, Kaminskyy D, Nektegayev I, Grellier P, Lesyk R. Isothiochromenothiazoles—A class of fused thiazolidinone derivatives with established anticancer activity that inhibits growth of *Trypanosoma brucei brucei*. Scientia Pharmaceutica. 2018;**86**(4):47. DOI: 10.3390/scipharm86040047

## Section 3

# Triazoles: Synthesis and Applications

Chapter 4

# 1,2,3-Triazoles: Synthesis and Biological Application

*Abdul Aziz Ali*

## Abstract

Among nitrogen-containing heterocyclic compounds, 1,2,3-triazoles are privileged structure motif and received a great deal of attention in academics and industry. Even though absent in nature, 1,2,3-triazoles have found broad applications in drug discovery, organic synthesis, polymer chemistry, supramolecular chemistry, bioconjugation, chemical biology, fluorescent imaging, and materials science. Therefore, the development of facile and straightforward methodology for the synthesis of 1,2,3-triazoles is of noteworthy interest. In this study, emphasis will be given to numerous synthetic approaches for the synthesis of 1,2,3-triazoles, especially the popular click chemistry approach. Furthermore, several biological activities of this promising heterocycle will also be discussed.

**Keywords:** 1,2,3-triazoles, click chemistry, organocatalysis, biological activity, drug discovery

## 1. Introduction

Nitrogen-containing heterocyclic compounds are indispensable for life as they are part of essential building blocks like amino acids, nucleotides, etc. 1,2,3-Triazoles are one of the most important nitrogen-containing five-membered heterocycles and have a wide range of applications in pharmaceuticals, supramolecular chemistry, organic synthesis, chemical biology and industry [1–6]. The 1,2,3-triazoles has numerous useful properties like high chemical stability (usually inert to acidic or basic hydrolysis as well as oxidizing and reducing conditions even at high temperature), aromatic character, strong dipole moment (4.8–5.6 Debye), and hydrogen bonding ability [7]. These spectacular features make the substituted 1,2,3-triazole motif structurally resembling to the amide bond, mimicking an E or a Z amide bond. Many prominent medicinal compounds having a 1,2,3-triazole core are available in the market like anticonvulsant drug Rufinamide, broad spectrum cephalosporin antibiotic cefatrizine, an anticancer drug carboxyamidotriazole and β-lactum antibiotic tazobactam, etc. [8].

## 2. Synthesis of 1,2,3-triazoles

Owing to its versatile applications, the synthesis of 1,2,3-triazoles has been a subject of extensive research. The synthetic methodologies for the preparation of this important scaffold can be broadly divided into four categories (**Figure 1**) [9]:

i. Huisgen 1,3-dipolar cycloaddition

ii. Metal-catalyzed 1,3-dipolar cycloaddition

iii. Strain-promoted azide alkyne cycloaddition

iv. Metal-free synthesis of 1,2,3-triazoles

## 2.1 Huisgen 1,3-dipolar cycloaddition

Huisgen 1,3-dipolar cycloaddition was the most straightforward and atom-economical synthesis of 1,2,3-triazoles. However, elevated reaction temperature and poor regioselectivity (mixtures of 1,4- and 1,5-isomers) make this process unsatisfactory [10].

## 2.2 Metal-catalyzed 1,3-dipolar cycloaddition

In 2001, Sharpless et al. coined the term "Click Chemistry," a set of highly reliable, practical, and selective reactions for the rapid synthesis of valuable new compounds and combinatorial libraries. The click reaction should be *modular, with*

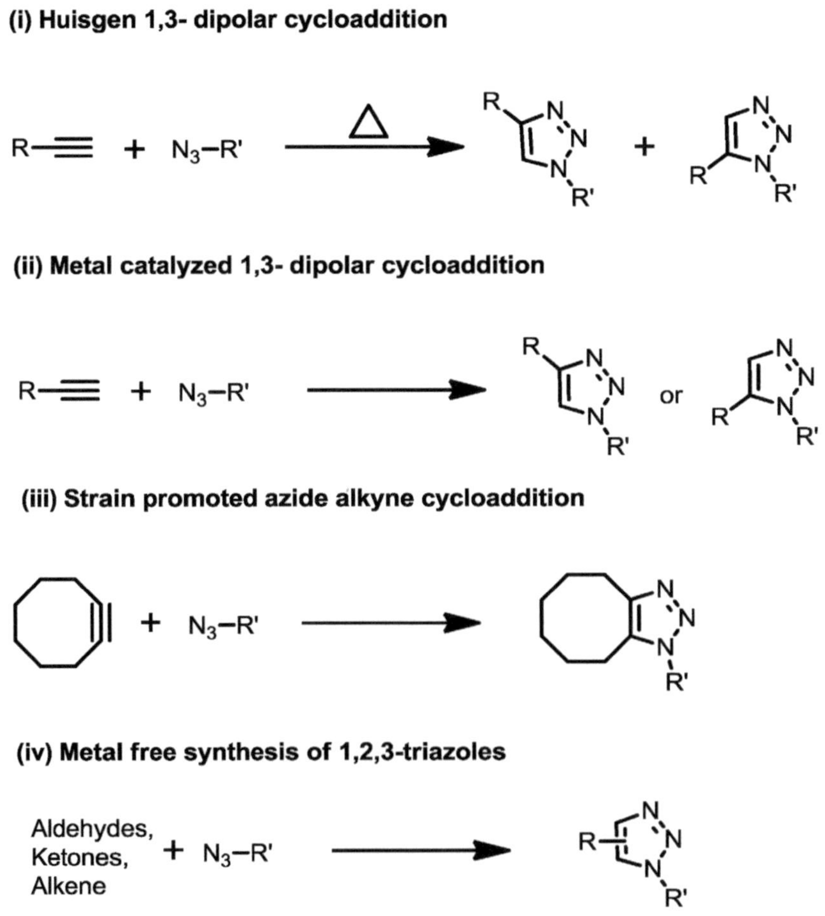


**Figure 1.**
*Strategy of the synthesis of 1,2,3-triazoles.*

*high yield, wide in scope, generate only innocuous by-products (that can be removed without chromatography), stereospecific, easy to carry out and that need benign solvent* [11]. In 2002, the groups of Sharpless and Meldal independently revealed a copper-catalyzed variant of Huisgen's azide-alkyne cycloaddition (CuAAC reaction) identified as one of the prime example of click chemistry in the literature [12, 13]. The unique advantages of CuAAC reaction are excellent substrate scope, prominent atom economy, good regioselectivity (only 1,4-isomer), high yield of products and mild reaction conditions [14–17].

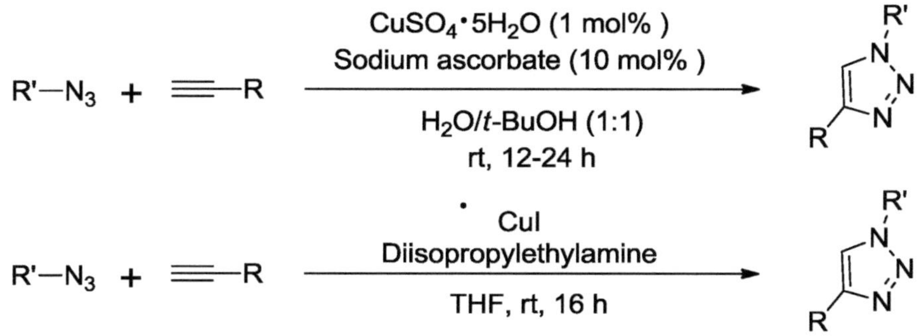


In 2005, Fokin and coworkers devised an efficient approach for the construction of 1,5-disubstituted 1,2,3-triazoles by ruthenium cyclopentadienyl complexes (RuAAC). In addition, internal alkynes also effective in this protocol leading to fully substituted 1,2,3-triazoles [18].

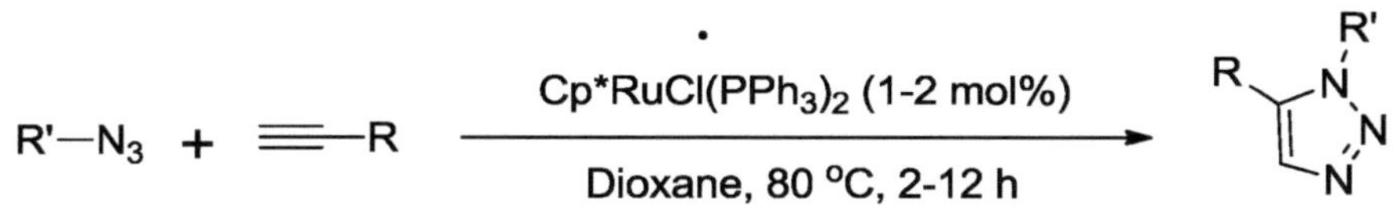


The McNulty group reported a well-defined Ag(I) complex for the regioselective synthesis of 1,4-disubstituted 1,2,3-triazoles at room temperature [19].

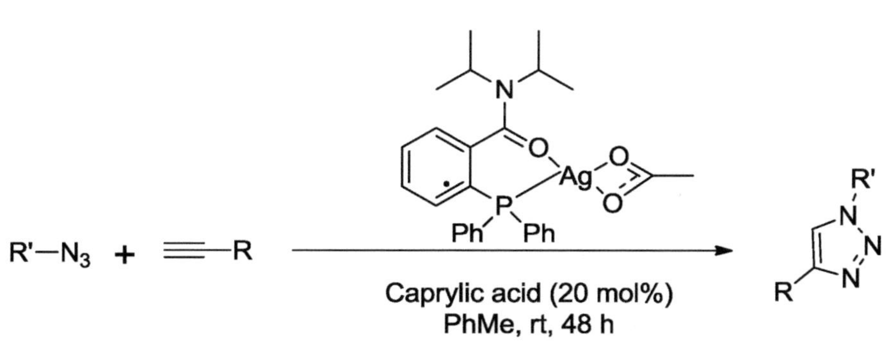


An interesting Zn(OAc)$_2$-catalyzed azide-alkyne cycloaddition was developed by Postnikov and his research group affording 1,4-disubstituted 1,2,3-triazoles [20].

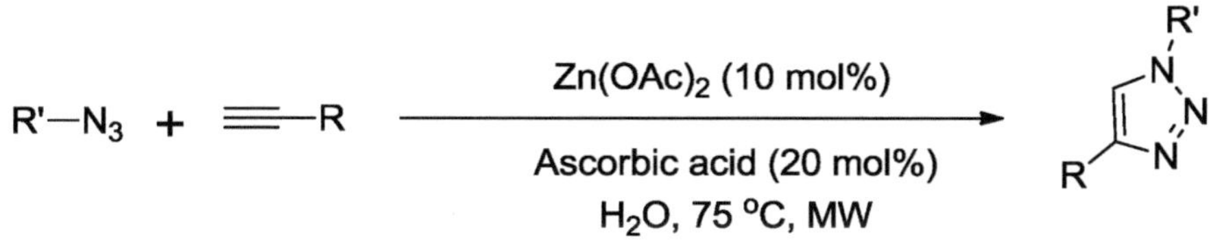

In 2017, Kim et al. devised $Cp_2Ni$/Xantphos catalytic method to access 1,5-disubstituted 1,2,3-triazoles under mild condition [21].

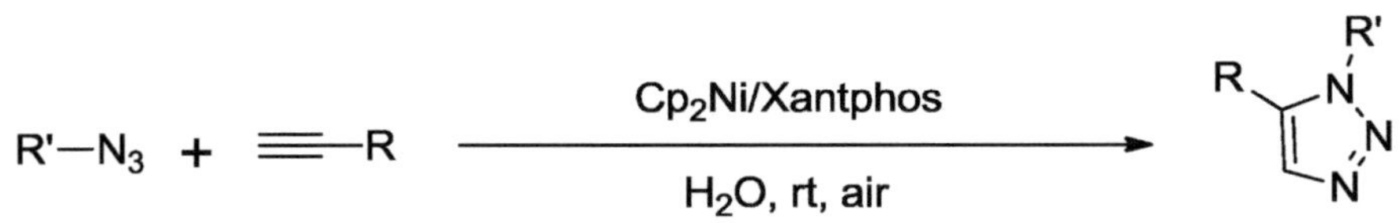


Sun and coworkers reported intermolecular iridium-catalyzed azide-alkyne cycloaddition reaction (IrAAC) of electron-rich internal alkynes [22].

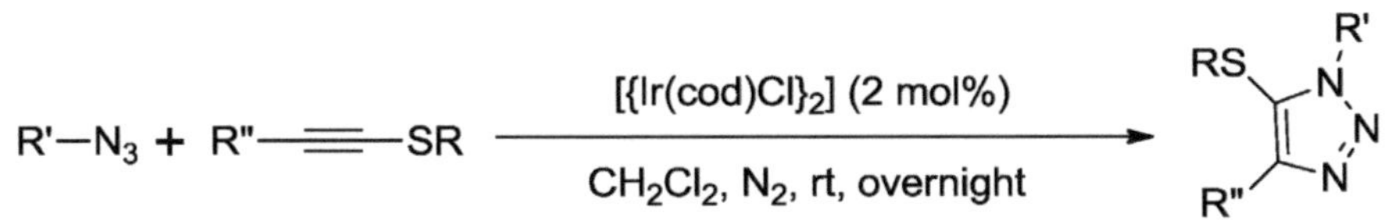


## 2.3 Strain-promoted azide alkyne cycloaddition

Despite the overwhelming popularity of click chemistry in modern science and technology, the using of metals creates serious concern in biological system due to cellular toxicity. The Bertozzi group explored an interesting protocol of strain-promoted azide-alkyne cycloaddition (SPAAC) reaction for bioconjugation. The driving force for this reaction was the release of large ring strain in the cycloalkynes which proceeds under physiological condition without any catalyst [23].

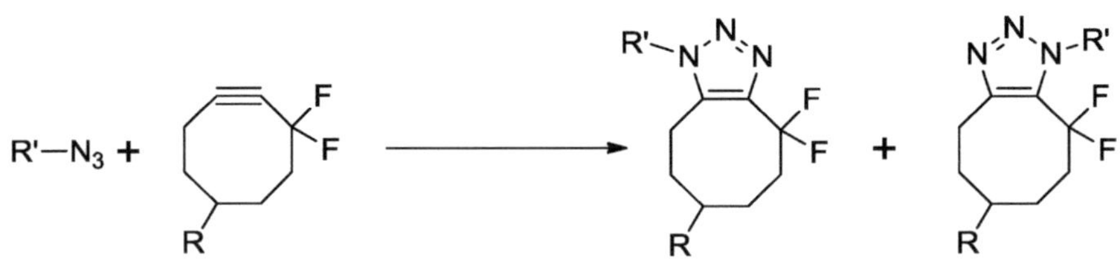


## 2.4 Metal free synthesis of 1,2,3-triazoles

Organocatalytic reactions has gained considerable attention in the synthesis of 1,2,3-triazoles using enamines, enolates as dipolarophiles. Besides, activated alkenes were established as a useful substrate for triazole formation.

Ramachary and coworkers developed L-proline-catalyzed synthesis of 1,2,3-triazoles via an enamine mediated [3 + 2]-cycloaddition reaction [24].

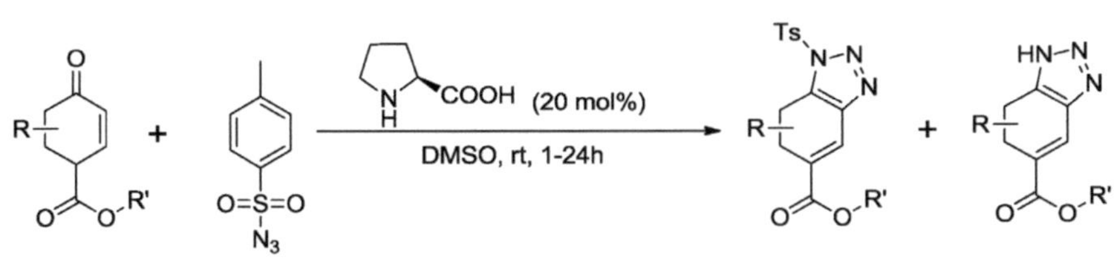

In 2011, the regioselective synthesis of 1,4,5-trisubstituted 1,2,3-triazoles was achieved by Wang et al. using an organocatalytic enamine azide reaction [25].

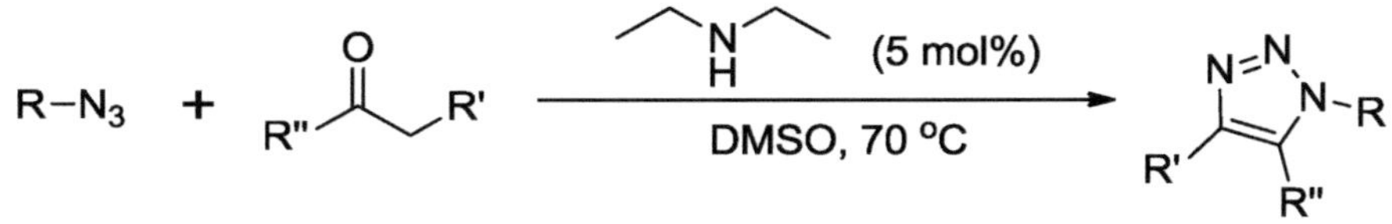


The Bressy group reported synthesis of substituted 1,2,3-triazoles from unactivated ketone and aromatic azide using microwave condition [26].

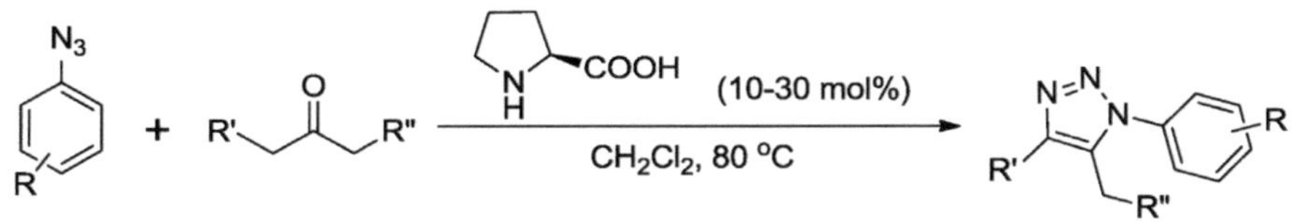


Wang and coworkers devised an organocatalytic method for the preparation of fully substituted 1,2,3-triazoles by diethylamine-catalyzed reaction of azides and allyl ketones [27].

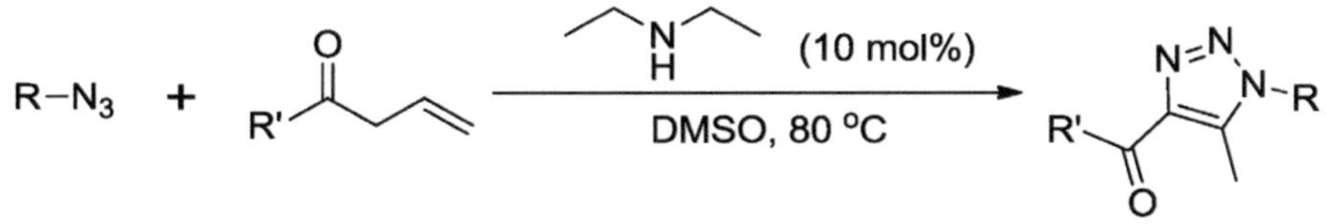


Iodine mediated, oxidant free synthesis of 1,5-disubstituted 1,2,3-triazoles was reported by the Wan group using primary amines, enamines and tosylhydrazine [28].

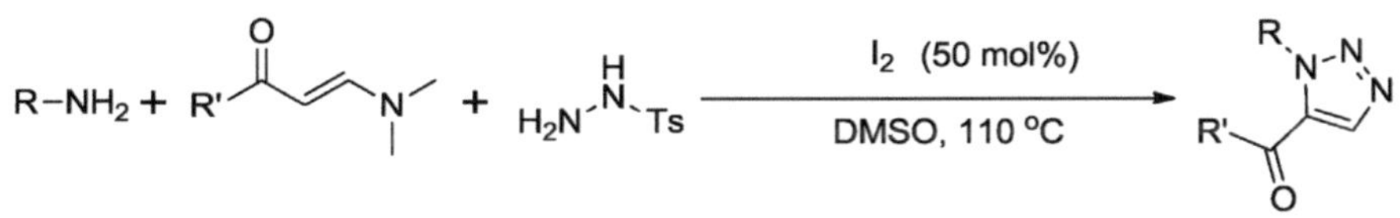


Using potassium carbonate, Kannan and co-workers developed a protocol for the synthesis of 4-acetyl-5-methyl-1,2,3-triazoles from acetylacetone and aromatic azides [29].

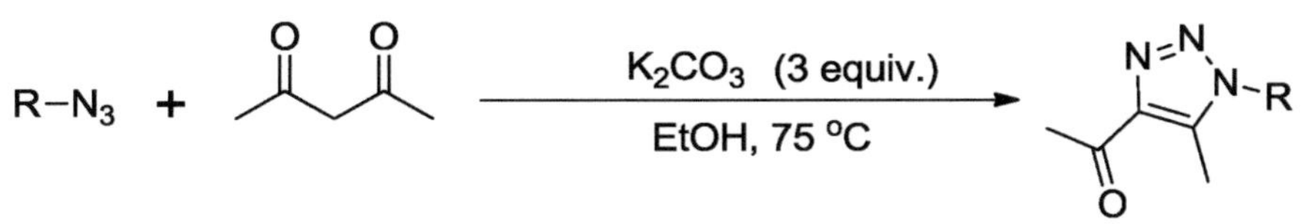


The Ramachary group described an efficient methodology for the preparation of 1,4-disubstituted 1,2,3-triazoles using organocatalytic azide-aldehyde [3 + 2] cycloaddition reaction [30].

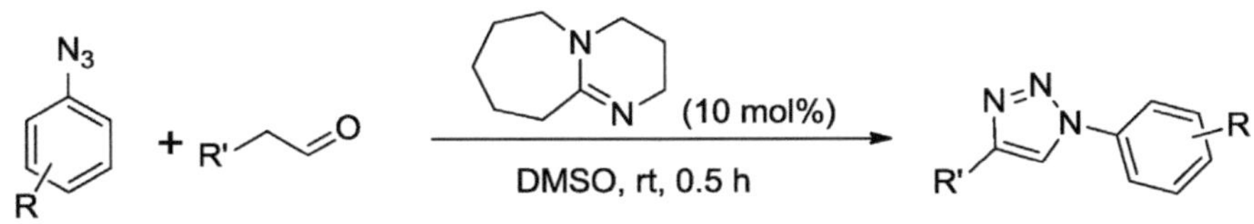

Paixão et al. reported the use of alkylidenemalononitriles in 1,3-dipolar cycloaddition with aromatic azides mediated by DBU [31].

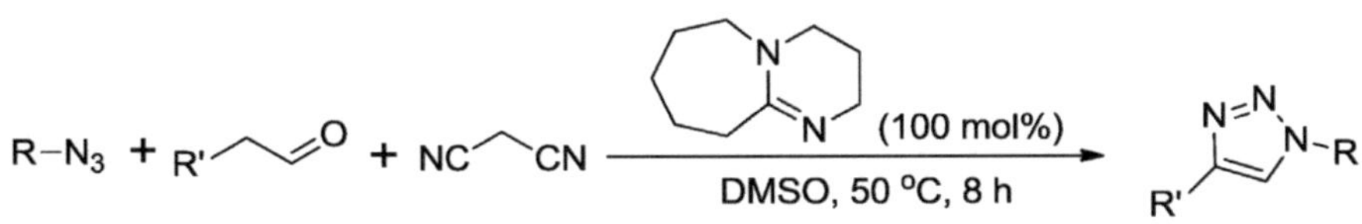

In their another pioneering work, Ramachary and coworkers reported an interesting organocatalytic [3 + 2]-cycloaddition reaction of ketones with azides for synthesis of fully substituted 1,2,3-triazoles [32].

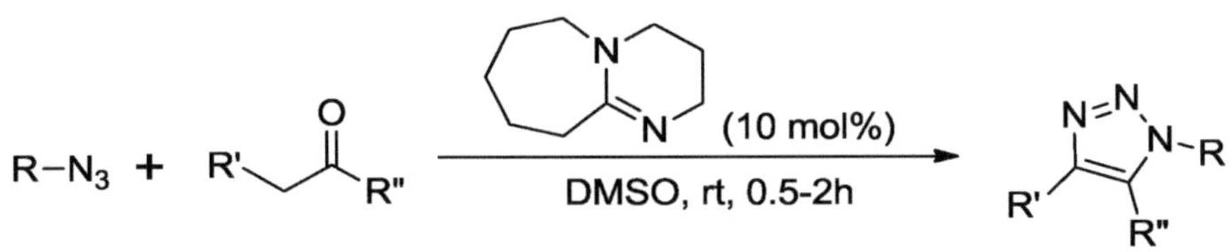

In a methodology published in 1986, Sakai et al. used primary amines and α,α-dichloro ketone derived tosylhydrazones for the metal free synthesis of 1,2,3-triazoles [33].

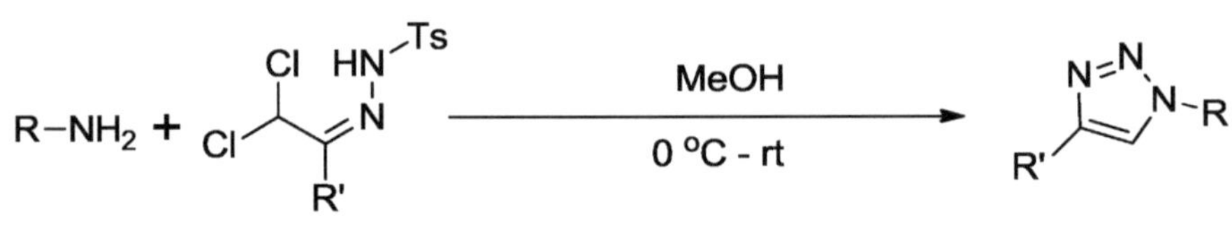

Westermann and co-workers developed a cascade reaction using α,α-dichloro-tosylhydrazones and primary amines in the presence of diisopropylethylamine [34].

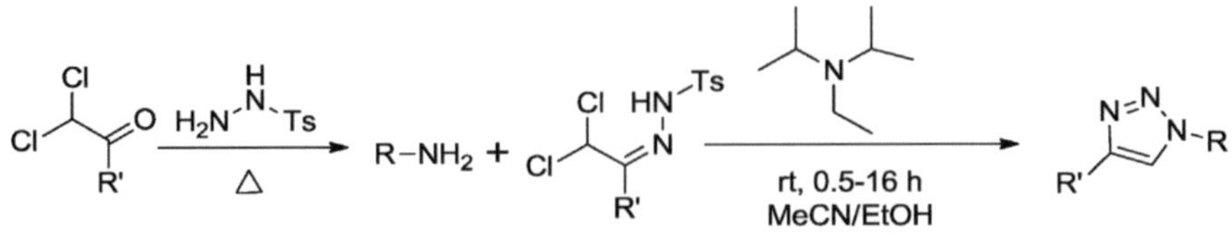

Metal free regioselective synthesis of 1,4,5-trisubstituted 1,2,3-triazoles was reported by Dehaen et al. from aldehydes, nitroalkanes and organic azides [35].

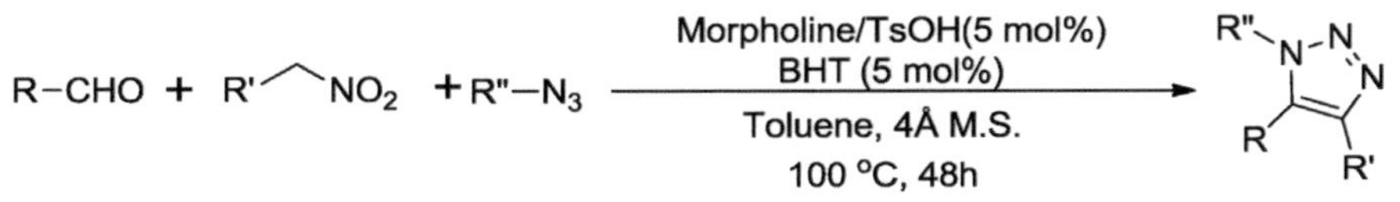


The Guan group developed p-toluenesulfonic acid-catalyzed 1,3-dipolar cycloaddition reaction for the synthesis of 4-aryl-NH-1,2,3-triazoles from nitroolefins with sodium azide [36].

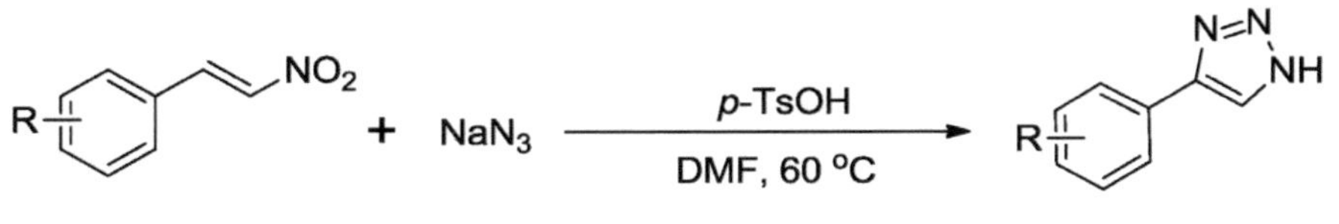


## 3. Biological activity of 1,2,3-triazoles

1,2,3-triazoles are stable towards metabolic degradation and easily form hydrogen bonding which can increase solubility favoring the binding of biomolecular targets. Owing to their unique properties, 1,2,3-triazoles are attractive building blocks in drug discovery.

### 3.1 Anti-cancer activity

Cancer is a major public health concern and second leading cause of mortality globally. Despite that numerous anticancer agents including taxol, vincristine, vinblastine, camptothecin derivatives, topotecan are available, search for novel compounds with different modes of actions has received significant interest.

Kallander et al. reported 4-aryl-1,2,3-triazoles **1** as inhibitors of human methionine aminopeptidase type 2 (hMetAP2). The anticancer activity of these molecules is due to the N1 and N2 nitrogen atoms of the triazole moiety that actively contribute in binding to the active site of enzyme [37].

Odlo and coworkers disclosed a series of cis-restricted 1,5-disubstituted 1,2,3-triazole analogues of combretastatin A-4. One of the triazole derivatives **2** showed effective cytotoxic activity against various cancer cell lines with $IC_{50}$ values in the nanomolar range. Molecular docking study shows that the triazole moiety interacts with β-tubulin via H-bonding with numerous amino acids [38].

The series of triazole-modified 20,30-dideoxy-20,30-diethanethioribonucleosides **3** displayed considerably better antitumor activity towards HepG2, A549, and Hela cell lines and higher cytotoxicity towards HepG2, LAC, and Hela cell lines compared to the control drug floxuridine [39].

Rangappa and coworkers prepared a series of 1,2-benzisoxazole tethered 1,2,3-triazoles **4** and established its noteworthy antiproliferative effect against human acute myeloid leukemia (AML) cells. Using MTT assay, 3-(4-(4-phenoxyphenyl)-1H-1,2,3-triazol-1-yl)benzo[d]isoxazole was found to be the most potent antiproliferative agent with an $IC_{50}$ of 2 μM against MV4-11 cells [40].

Using "click chemistry" approach, the Miller group prepared a series of N-((1-benzyl-1H-1,2,3-triazol-4-yl)methyl)arylamides and examined their antiproliferative activity. One of the compound **5** displayed an $IC_{50}$ of 46 nM against MCF-7 human breast tumor cells [41].

Lin and coworkers synthesized a series of heterocycle-fused 1,2,3-triazoles and evaluated their cytotoxic activity. With $IC_{50}$ values lower than 1.9 μg/mL against A431 and K562 human tumor cell lines, 4-Methoxyphenyl substituted 1,3-oxazoheterocycle fused 1,2,3-triazole **6** was found to be the most potent derivative [42].

1,2,3-triazole derivatives of betulinic acid were synthesized by Koul et al. and their cytotoxic activity against nine human cancer cell lines was evaluated (**Figure 2**). Two molecules 7 and **8** exhibited notable $IC_{50}$ values (2.5 and 3.5 μM, respectively) against leukemia cell line HL-60 (5–7-fold higher potency than betulinic acid) [43].

### 3.2 Anti-inflammatory activity

Inflammation is particularly complex biological process of body tissues, where membrane-bound phospholipids release arachidonic acid (AA), followed by

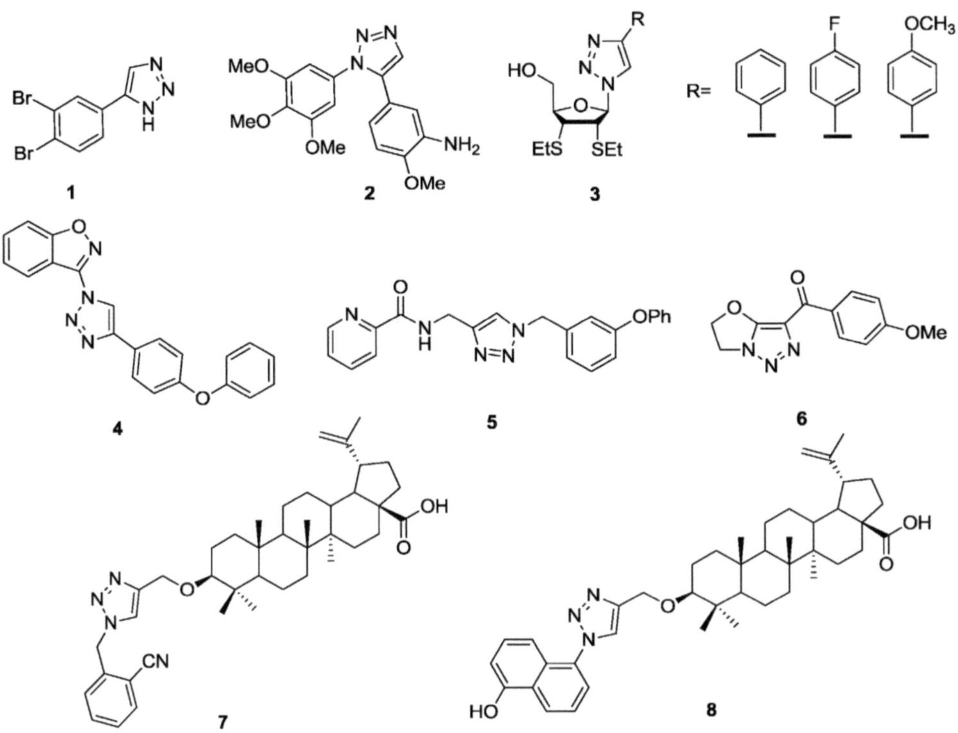

**Figure 2.**
*Some examples of 1,2,3-triazole containing molecules with anticancer activity.*

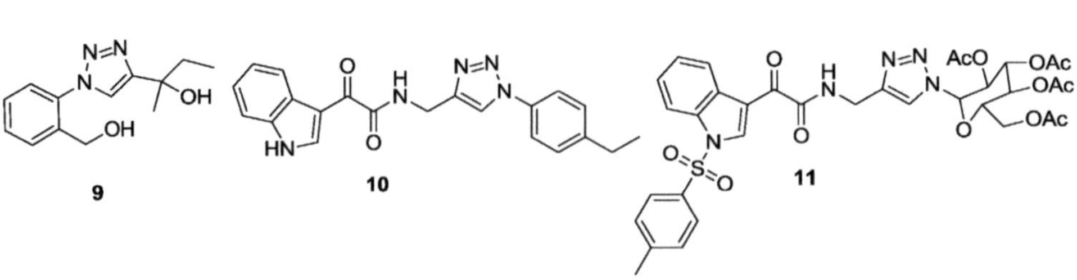

**Figure 3.**
*Various examples of 1,2,3-triazole containing molecules with anti-inflammatory activity.*

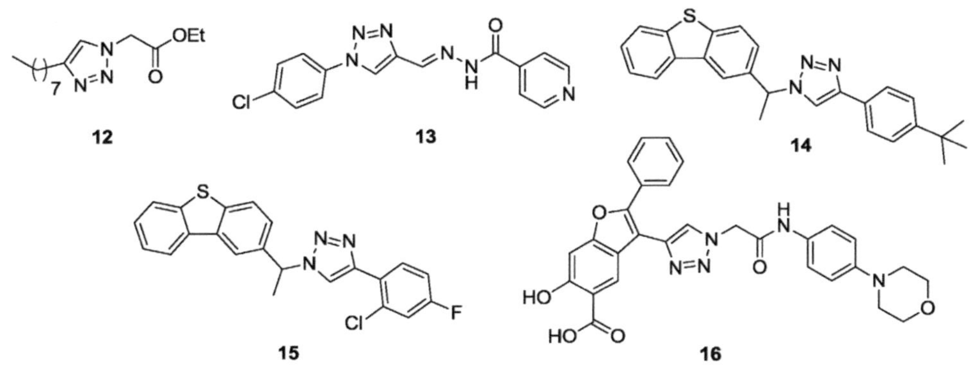

**Figure 4.**
*Representative examples of 1,2,3-triazole containing molecules with antitubercular activity.*

biotransformation processes using cycloxygenase (COX) and 5-lipoxygenase (5-LOX) pathways. Several non-steroidal anti-inflammatory drugs (NSAIDs) such as indomethacin, ibuprofen, and naproxen block arachidonic acid metabolism by obstructing cycloxygenase. Nevertheless the side effects associated with these drugs prompted medicinal chemists to develop alternative scaffolds.

The Jung group synthesized twenty-four phenyl-1H-1,2,3-triazole derivatives and studied their biological activity. At the same dose of 25 mg/kg, compound **9** showed more compelling effects than the existing anti-inflammatory drug diclofenac [44].

Yar and coworkers reported 1,2,3-triazole tethered Indole-3-glyoxamide derivatives for in vivo anti-inflammatory activity using click chemistry approach. Two compounds **10** and **11** displayed excellent inhibition of COX-2 ($IC_{50}$ 0.12 μM) with good COX-2 selectivity index (COX-2/COX-1) of 0.058 and 0.046, respectively (**Figure 3**) [45].

### 3.3 Antitubercular activity

Tuberculosis (TB) caused by *Mycobacterium tuberculosis* is one of the infectious contagious disease and remains a serious risk to public health worldwide. Generally, the direct observed therapy strategy (DOTS) is the treatment for TB, but the emergence of multidrug-resistant TB (MDR-TB) and extensively drug-resistant TB (XDR-TB) developed challenges. Therefore identifying of effective anti-TB drug candidates has received enormous interest.

Labadie and coworkers used click chemistry to synthesize a small library of 1,2,3-triazole derivatives and screened them against *Mycobacterium tuberculosis* and *Mycobacterium avium*. The biological screening indicated that the triazole **12** displayed more significant activity against *M. tuberculosis* than standard drug [46].

Using click chemistry, the Boechat group reported 4-substituted N-phenyl-1,2,3-triazole derivatives for antimicrobial activity against *Mycobacterium tuberculosis* strain H37Rv (ATCC 27294). Derivatives of isoniazid, (E)-N′-[(1-aryl)-1H-1,2,3-triazole-4-yl)methylene] isonicotinoyl hydrazides, **13** revealed significant activity with minimum inhibitory concentration (MIC) value of 0.62 μg/mL [47].

The Kantevari group described a molecular hybridization approach for the synthesis of triazole clubbed dibenzo[b,d]thiophene-based *Mycobacterium tuberculosis* inhibitors. The most potent compounds **14** and **15** in check of their *in vitro* activity against *M. tuberculosis* strain H37Rv exhibited MIC = 0.78 μg/mL [48].

Zhang et al. synthesized triazole-based library of benzofuran salicylic acid derivatives using click chemistry strategy. The compound **16** was found to be potent antiTB therapeutic with efficient cellular activity (**Figure 4**) [49].

### 3.4 Antimicrobial activity

Fungal and bacterial infections create severe apprehension for human and animal survival. The inefficacy of available drugs and rising resistant strains demand significant interest into new classes of antimicrobial agents.

Agarwal and coworkers synthesized 1,2,3-triazole derivatives of chalcones and flavones by click chemistry and screened their antimicrobial and antiplasmodial activity. Several compound including **17** showed promising antifungal and antibacterial activity [50].

The Murugulla group studied antimicrobial activity of theophylline containing 1,2,3-triazoles with variant nucleoside derivatives. Compound **18** was shown to be potent and effective against three bacterial strains *B. cereus*, *Escherichia coli* and *P. aureoginosa* with MIC values of 0.0156, 0.03125, 0.0625 mg/mL and compound **19** with MIC values of 0.03125, 0.0156, 0.0625 mg/mL was found to be effective against *S. aureus, B. cereus* and *Escherichia coli*, respectively [51].

Diaryl sulfone containing novel 1,2,3-triazoles were synthesized by Jørgensen and coworkers and their biological evaluation was carried out as well. Compound **20** was found to be the most potent antifungal agents with MIC at 25 μg/mL [52].

Zhou et al. reported a series of 1,2,3-triazole-derived naphthalimides for potential antimicrobial activity. Bioactive assay revealed that **21** showed better anti-*Escherichia coli* activity than existing drugs Norfloxacin and Chloromycin [53].

5-nitrofuran—triazole congener—was prepared by the Kamal group and its biological activity was studied. Among the other compounds, **22** exhibited promising antibacterial activity (MIC value of 1.9 μg/mL against different bacterial strains)

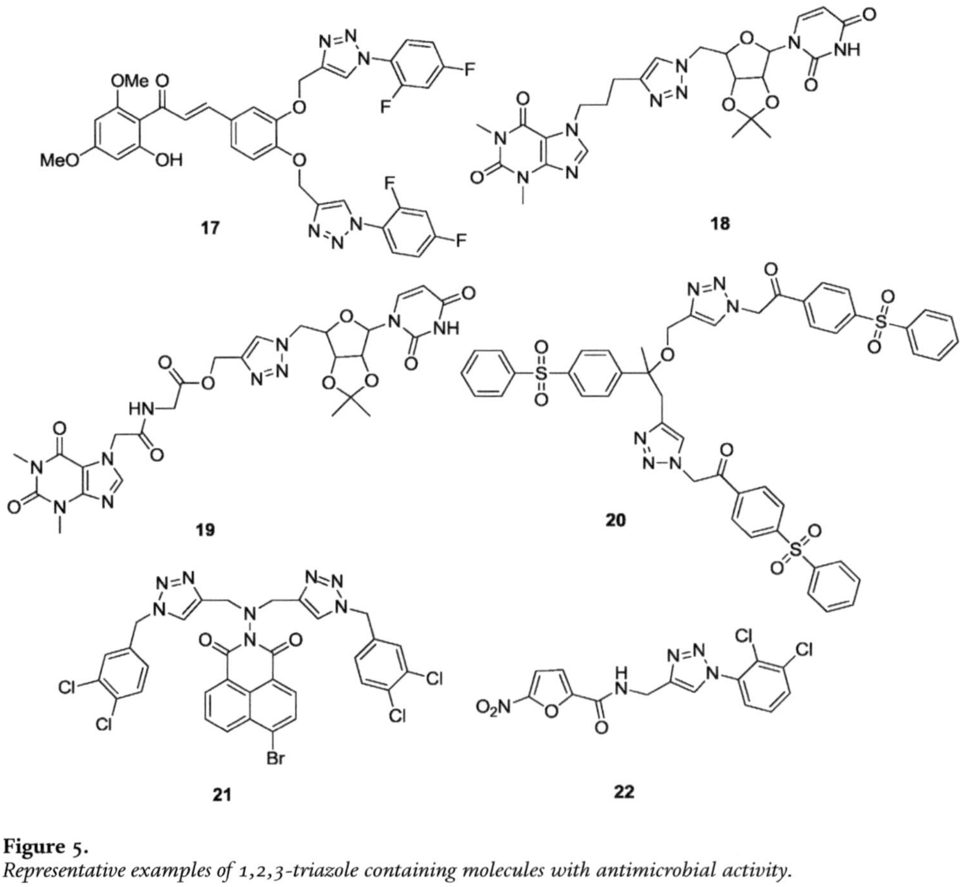

**Figure 5.**
*Representative examples of 1,2,3-triazole containing molecules with antimicrobial activity.*

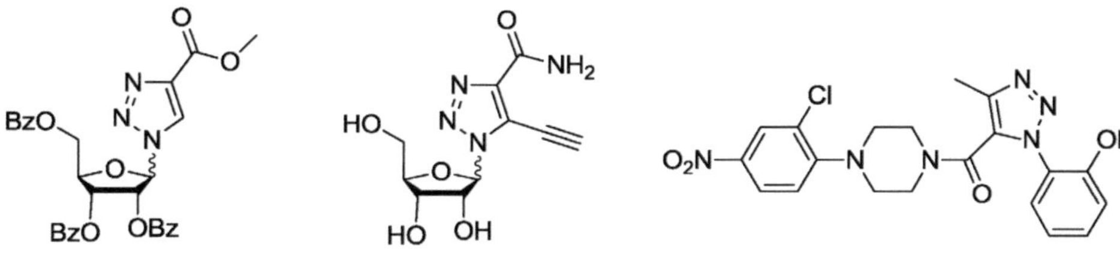


**Figure 6.**
*Examples of 1,2,3-triazole containing molecules with antiviral activity.*

and antifungal activity (MIC = 3.9 μg/mL) compared to the standard miconazole (MIC = 7.8 μg/mL) against *C. albicans* and *C. parapsilosis* (**Figure 5**) [54].

### 3.5 Antiviral activity

Viral diseases are caused by viruses infecting an organism body. Although vaccines and antiviral drugs are used for treating viral infections, advance of novel viruses creates health risk over the world. Therefore development of alternative antiviral agents is of significant interest.

Boechat and coworkers reported the synthesis of 1,2,3-triazole nucleoside ribavirin analogs and studied their antiviral activity. The synthesized compound **23** displayed potent activity with $IC_{50}$ values 14 and 3.8 μM for Influenza A and reverse transcriptase (RT) from human immunodeficiency virus type 1 (HIV-1 RT), respectively [55].

Ribavirin analogues—4,5-disubstituted 1,2,3-triazole nucleosides—were synthesized by Zeidler et al. and screened for their biological activity. 5-ethynyl nucleoside **24** exhibited effective virus-inhibitory activity against influenza A (H1N1, H3N2 and H5N1), influenza B, measles and respiratory syncytial viruses [56].

The Ding group targeted virus nucleoprotein and synthesized 1,2,3-triazole-4-carboxamide derivatives for anti-influenza drug development. The compound **25**, inhibited the replication of various H3N2 and H1N1 influenza A virus strains with $IC_{50}$ values ranging from 0.5 to 4.6 μM (**Figure 6**) [57].

## 4. Conclusion

In summary, 1,2,3-triazole moiety has proven to be a privileged scaffolds in medicinal chemistry. The exceptional properties of this promising heterocycle facilitate its wide range of applications from material science to bioconjugation. Thanks to Sharpless for introducing "Click Chemistry," one of the most prevailing tools in drug discovery, chemical biology, and proteomic applications and undoubtedly opens new avenue to the scientific community towards the improvement of life.

## Acknowledgements

The author is thankful for the financial support by CSIR, New Delhi, India.

## Conflict of interest

There are no conflicts to declare.

## Author details

Abdul Aziz Ali
Material Science and Technology Division, CSIR-North East Institute of Science and Technology, Jorhat, Assam, India

*Address all correspondence to: aaziz496@gmail.com

## References

[1] Dheer D, Singh V, Shankar R. Medicinal attributes of 1,2,3-triazoles: Current developments. Bioorganic Chemistry. 2017;**71**:30-54. DOI: 10.1016/j.bioorg.2017.01.010

[2] Schulze B, Schubert US. Beyond click chemistry–Supramolecular interactions of 1,2,3-triazoles. Chemical Society Reviews. 2014;**43**:2522-2571. DOI: 10.1039/C3CS60386E

[3] Liang L, Astruc D. The copper (I)-catalyzed alkyne-azide cycloaddition (CuAAC)"click" reaction and its applications. An overview. Coordination Chemistry Reviews. 2011;**255**: 2933-2945. DOI: 10.1016/j.ccr.2011.06.028

[4] Ngo JT, Adams SR, Deerinck TJ, Boassa D, Rodriguez-Rivera F, Palida SF, et al. Click-EM for imaging metabolically tagged nonprotein biomolecules. Nature Chemical Biology. 2016;**12**:459-465. DOI: 10.1038/nchembio.2076

[5] Astruc D, Liang L, Rapakousiou A, Ruiz J. Click dendrimers and triazole-related aspects: Catalysts, mechanism, synthesis, and functions. A bridge between dendritic architectures and nanomaterials. Accounts of Chemical Research. 2012;**45**:630-640. DOI: 10.1021/ar200235m

[6] Wang X, Huang B, Liu X, Zhan P. Discovery of bioactive molecules from CuAAC click-chemistry-based combinatorial libraries. Drug Discovery Today. 2016;**21**:118-132. DOI: 10.1016/j.drudis.2015.08.004

[7] Meldal M, Tornøe CW. Cu-catalyzed azide—alkyne cycloaddition. Chemical Reviews. 2008;**108**:2952-3015. DOI: 10.1021/cr0783479

[8] Agalave SG, Maujan SR, Pore VS. Click chemistry: 1,2,3-Triazoles as pharmacophores. Chemistry – An Asian Journal. 2011;**6**:2696-2718. DOI: 10.1002/asia.201100432

[9] Jalani HB, Karagöz AÇ, Tsogoeva SB. Synthesis of substituted 1,2,3-triazoles via metal-free click cycloaddition reactions and alternative cyclization methods. Synthesis. 2017;**49**:29-41. DOI: 10.1055/s-0036-1588904

[10] Huisgen R. 1,3-Dipolar Cycloadditions. Proceedings of the Chemical Society. 1961:357. DOI: 10.1039/PS9610000357

[11] Kolb HC, Finn MG, Sharpless KB. Click chemistry: Diverse chemical function from a few good reactions. Angewandte Chemie International Edition. 2001;**40**:2004-2021. DOI: 10.1002/1521-3773(20010601)40:11<2004::AID-ANIE2004>3.0.CO;2-5

[12] Rostovtsev VV, Green LG, Fokin VV, Sharpless KB. Angewandte Chemie International Edition. 2002;**41**: 2596-2599. DOI: 10.1002/1521-3773(20020715)41:14<2596::AID-ANIE2596>3.0.CO;2-4

[13] Tornøe CW, Christensen C, Meldal M. Peptidotriazoles on solid phase: [1,2,3]-Triazoles by regiospecific copper (I)-catalyzed 1, 3-dipolar cycloadditions of terminal alkynes to azides. The Journal of Organic Chemistry. 2002;**67**:3057-3064. DOI: 10.1021/jo011148j

[14] Wang Q, Chan TR, Hilgraf R, Fokin VV, Sharpless KB, Finn MG. Bioconjugation by copper (I)-catalyzed azide-alkyne [3+ 2] cycloaddition. Journal of the American Chemical Society. 2003;**125**:3192-3193. DOI: 10.1021/ja021381e

[15] Wu P, Feldman AK, Nugent AK, Hawker CJ, Scheel A, Voit B, et al. Efficiency and fidelity in a

click-chemistry route to triazole dendrimers by the copper (I)-catalyzed ligation of azides and alkynes. Angewandte Chemie International Edition. 2004;**43**:3928-3932. DOI: 10.1002/anie.200454078

[16] Chan TR, Hilgraf R, Sharpless KB, Fokin VV. Polytriazoles as copper (I)-stabilizing ligands in catalysis. Organic Letters. 2004;**6**:2853-2855. DOI: 10.1021/ol0493094

[17] Himo F, Lovell T, Hilgraf R, Rostovtsev VV, Noodleman L, Sharpless KB, et al. Copper (I)-catalyzed synthesis of azoles. DFT study predicts unprecedented reactivity and intermediates. Journal of the American Chemical Society. 2005;**127**:210-216. DOI: 10.1021/ja0471525

[18] Zhang L, Chen X, Xue P, Sun HH, Williams ID, Sharpless KB, et al. Ruthenium-catalyzed cycloaddition of alkynes and organic azides. Journal of the American Chemical Society. 2005;**127**: 15998-15999. DOI: 10.1021/ja054114s

[19] McNulty J, Keskar K, Vemula R. The first well-defined silver (I)-complex-catalyzed cycloaddition of azides onto terminal alkynes at room temperature. Chemistry – A European Journal. 2011; **17**:14727-14730. DOI: 10.1002/chem.201103244

[20] Morozova MA, Yusubov MS, Kratochvil B, Eigner V, Bondarev AA, Yoshimura A, et al. Regioselective Zn $(OAc)_2$-catalyzed azide–alkyne cycloaddition in water: The green click-chemistry. Organic Chemistry Frontiers. 2017;**4**:978-985. DOI: 10.1039/C6QO00787B

[21] Kim WG, Kang ME, Lee JB, Jeon MH, Lee S, Lee J, et al. Nickel-catalyzed azide–alkyne cycloaddition to access 1, 5-disubstituted 1,2,3-triazoles in air and water. Journal of the American Chemical Society. 2017;**139**: 12121-12124. DOI: 10.1021/jacs.7b06338

[22] Ding S, Jia G, Sun J. Iridium-catalyzed intermolecular azide–alkyne cycloaddition of internal thioalkynes under mild conditions. Angewandte Chemie International Edition. 2014;**53**: 1877-1880. DOI: 10.1002/anie.201309855

[23] Codelli JA, Baskin JM, Agard NJ, Bertozzi CR. Second-generation difluorinated cyclooctynes for copper-free click chemistry. Journal of the American Chemical Society. 2008;**130**: 11486-11493. DOI: 10.1021/ja803086r

[24] Ramachary DB, Ramakumar K, Narayana VV. Amino acid-catalyzed cascade [3+ 2]-cycloaddition/hydrolysis reactions based on the push–pull dienamine platform: Synthesis of highly functionalized NH-1,2,3-triazoles. Chemistry – A European Journal. 2008; **14**:9143-9147. DOI: 10.1002/chem.200801325

[25] Danence LJT, Gao Y, Li M, Huang Y, Wang J. Organocatalytic enamide–azide cycloaddition reactions: Regiospecific synthesis of 1, 4, 5-trisubstituted-1,2,3-triazoles. Chemistry – A European Journal. 2011;**17**:3584-3587. DOI: 10.1002/chem.201002775

[26] Belkheira M, El Abed D, Pons JM, Bressy C. Organocatalytic synthesis of 1,2,3-triazoles from unactivated ketones and arylazides. Chemistry – A European Journal. 2011;**17**:12917-12921. DOI: 10.1002/chem.201102046

[27] Li W, Du Z, Huang J, Jia Q, Zhang K, Wang J. Direct access to 1,2,3-triazoles through organocatalytic 1, 3-dipolar cycloaddition reaction of allyl ketones with azides. Green Chemistry. 2014;**16**:3003-3006. DOI: 10.1039/C4GC00406J

[28] Wan JP, Cao S, Liu Y. A metal-and azide-free multicomponent assembly toward regioselective construction of 1, 5-disubstituted 1,2,3-triazoles. The Journal of Organic Chemistry. 2015;**80**:

9028-9033. DOI: 10.1021/acs.joc.5b01121

[29] Kamalraj VR, Senthil S, Kannan P. One-pot synthesis and the fluorescent behavior of 4-acetyl-5-methyl-1,2,3-triazole regioisomers. Journal of Molecular Structure. 2008;**892**:210-215. DOI: 10.1016/j.molstruc.2008.05.028

[30] Ramachary DB, Shashank AB, Karthik S. An organocatalytic azide–aldehyde [3+ 2] cycloaddition: High-yielding regioselective synthesis of 1, 4-disubstituted 1,2,3-triazoles. Angewandte Chemie International Edition. 2014;**53**:10420-10424. DOI: 10.1002/anie.201406721

[31] Ali A, Corrêa AG, Alves D, Zukerman-Schpector J, Westermann B, Ferreira MA, et al. An efficient one-pot strategy for the highly regioselective metal-free synthesis of 1, 4-disubstituted-1,2,3-triazoles. Chemical Communications. 2014;**50**:11926-11929. DOI: 10.1039/C4CC04678A

[32] Shashank AB, Karthik S, Madhavachary R, Ramachary DB. An enolate-mediated organocatalytic azide–ketone [3+ 2]-cycloaddition reaction: Regioselective high-yielding synthesis of fully decorated 1,2,3-triazoles. Chemistry – A European Journal. 2014;**20**:16877-16881. DOI: 10.1002/chem.201405501

[33] Sakai K, Hida N, Kondo K. Reactions of α-polyhalo ketone tosylhydrazones with sulfide ion and primary amines. Cyclization to 1,2,3-thiadiazoles and 1,2,3-triazoles. Bulletin of the Chemical Society of Japan. 1986;**59**:179-183. DOI: 10.1246/bcsj.59.179

[34] van Berkel SS, Brauch S, Gabriel L, Henze M, Stark S, Vasilev D, et al. Traceless tosylhydrazone-based triazole formation: A metal-free alternative to strain-promoted azide–alkyne cycloaddition. Angewandte Chemie International Edition. 2012;**51**:5343-5346. DOI: 10.1002/anie.201108850

[35] Thomas J, John J, Parekh N, Dehaen W. A metal-free three-component reaction for the regioselective synthesis of 1, 4, 5-trisubstituted 1,2,3-triazoles. Angewandte Chemie International Edition. 2014;**53**:10155-10159. DOI: 10.1002/anie.201403453

[36] Quan XJ, Ren ZH, Wang YY, Guan ZH. p-toluenesulfonic acid mediated 1, 3-dipolar cycloaddition of nitroolefins with NaN3 for synthesis of 4-aryl-NH-1,2,3-triazoles. Organic Letters. 2014;**16**:5728-5731. DOI: 10.1021/ol5027975

[37] Kallander LS, Lu Q, Chen W, Tomaszek T, Yang G, Tew D, et al. 4-Aryl-1,2,3-triazole: A novel template for a reversible methionine aminopeptidase 2 inhibitor, optimized to inhibit angiogenesis in vivo. Journal of Medicinal Chemistry. 2005;**48**:5644-5647

[38] Odlo K, Hentzen J, dit Chabert JF, Ducki S, Gani OA, Sylte I, et al. 1, 5-Disubstituted 1,2,3-triazoles as cis-restricted analogues of combretastatin A-4: Synthesis, molecular modeling and evaluation as cytotoxic agents and inhibitors of tubulin. Bioorganic & Medicinal Chemistry. 2008;**16**:4829-4838. DOI: 10.1016/j.bmc.2008.03.049

[39] Yu JL, Wu QP, Zhang QS, Liu YH, Li YZ, Zhou ZM. Synthesis and antitumor activity of novel 2′, 3′-dideoxy-2′, 3′-diethanethionucleosides bearing 1,2,3-triazole residues. Bioorganic & Medicinal Chemistry Letters. 2010;**20**:240-243. DOI: 10.1016/j.bmcl.2009.10.127

[40] Ashwini N, Garg M, Mohan CD, Fuchs JE, Rangappa S, Anusha S, et al. Synthesis of 1, 2-benzisoxazole tethered 1,2,3-triazoles that exhibit anticancer

activity in acute myeloid leukemia cell lines by inhibiting histone deacetylases, and inducing p21 and tubulin acetylation. Bioorganic & Medicinal Chemistry. 2015;**23**:6157-6165. DOI: 10.1016/j.bmc.2015.07.069

[41] Stefely JA, Palchaudhuri R, Miller PA, Peterson RJ, Moraski GC, Hergenrother PJ, et al. N-((1-benzyl-1 H-1,2,3-triazol-4-yl) methyl) arylamide as a new scaffold that provides rapid access to antimicrotubule agents: Synthesis and evaluation of antiproliferative activity against select cancer cell lines. Journal of Medicinal Chemistry. 2010;**53**:3389-3395. DOI: 10.1021/jm1000979

[42] Yan SJ, Liu YJ, Chen YL, Liu L, Lin J. An efficient one-pot synthesis of heterocycle-fused 1,2,3-triazole derivatives as anti-cancer agents. Bioorganic & Medicinal Chemistry Letters. 2010;**20**:5225-5228. DOI: 10.1016/j.bmcl.2010.06.141

[43] Majeed R, Sangwan PL, Chinthakindi PK, Khan I, Dangroo NA, Thota N, et al. Synthesis of 3-O-propargylated betulinic acid and its 1,2,3-triazoles as potential apoptotic agents. European Journal of Medicinal Chemistry. 2013;**63**:782-792. DOI: 10.1016/j.ejmech.2013.03.028

[44] Kim TW, Yong Y, Shin SY, Jung H, Park KH, Lee YH, et al. Synthesis and biological evaluation of phenyl-1H-1,2,3-triazole derivatives as anti-inflammatory agents. Bioorganic Chemistry. 2015;**59**:1-11. DOI: 10.1016/j.bioorg.2015.01.003

[45] Naaz F, Pallavi MP, Shafi S, Mulakayala N, Yar MS, Kumar HS. 1,2,3-triazole tethered indole-3-glyoxamide derivatives as multiple inhibitors of 5-LOX, COX-2 & tubulin: Their anti-proliferative & anti-inflammatory activity. Bioorganic Chemistry. 2018;**81**:1-20. DOI: 10.1016/j.bioorg.2018.07.029

[46] Labadie GR, de la Iglesia A, Morbidoni HR. Targeting tuberculosis through a small focused library of 1,2,3-triazoles. Molecular Diversity. 2011;**15**: 1017-1024. DOI: 10.1007/s11030-011-9319-0

[47] Boechat N, Ferreira VF, Ferreira SB, Ferreira MLG, Silva FC, Bastos MM, et al. Novel 1,2,3-triazole derivatives for use against *Mycobacterium tuberculosis* H37Rv (ATCC 27294) strain. Journal of Medicinal Chemistry. 2011;**54**: 5988-5999. DOI: 10.1021/jm2003624

[48] Patpi SR, Pulipati L, Yogeeswari P, Sriram D, Jain N, Sridhar B, et al. Design, synthesis, and structure–activity correlations of novel dibenzo [b, d] furan, dibenzo [b, d] thiophene, and N-methylcarbazole clubbed 1,2,3-triazoles as potent inhibitors of mycobacterium tuberculosis. Journal of Medicinal Chemistry. 2012;**55**: 3911-3922. DOI: 10.1021/jm300125e

[49] Zhou B, He Y, Zhang X, Xu J, Luo Y, Wang Y, et al. Targeting mycobacterium protein tyrosine phosphatase B for antituberculosis agents. Proceedings of the National Academy of Sciences. 2010;**107**: 4573-4578. DOI: 10.1073/pnas.0909133107

[50] Kant R, Kumar D, Agarwal D, Gupta RD, Tilak R, Awasthi SK, et al. Synthesis of newer 1,2,3-triazole linked chalcone and flavone hybrid compounds and evaluation of their antimicrobial and cytotoxic activities. European Journal of Medicinal Chemistry. 2016; **113**:34-49. DOI: 10.1016/j.ejmech.2016.02.041

[51] Ruddarraju RR, Murugulla AC, Kotla R, Tirumalasetty MCB, Wudayagiri R, Donthabakthuni S, et al. Design, synthesis, anticancer, antimicrobial activities and molecular docking studies of theophylline containing acetylenes and theophylline

containing 1,2,3-triazoles with variant nucleoside derivatives. European Journal of Medicinal Chemistry. 2016; **123**:379-396. DOI: 10.1016/j.ejmech.2016.07.024

[52] Mady MF, Awad GE, Jørgensen KB. Ultrasound-assisted synthesis of novel 1,2,3-triazoles coupled diaryl sulfone moieties by the CuAAC reaction, and biological evaluation of them as antioxidant and antimicrobial agents. European Journal of Medicinal Chemistry. 2014;**84**:433-443. DOI: 10.1016/j.ejmech.2014.07.042

[53] Lv JS, Peng XM, Kishore B, Zhou CH. 1,2,3-Triazole-derived naphthalimides as a novel type of potential antimicrobial agents: Synthesis, antimicrobial activity, interaction with calf thymus DNA and human serum albumin. Bioorganic & Medicinal Chemistry Letters. 2014;**24**: 308-313. DOI: 10.1016/j.bmcl.2013.11.013

[54] Kamal A, Hussaini SA, Sucharitha ML, Poornachandra Y, Sultana F, Kumar CG. Synthesis and antimicrobial potential of nitrofuran–triazole congeners. Organic & Biomolecular Chemistry. 2015; **13**:9388-9397. DOI: 10.1039/C5OB01353D

[55] Maria de Lourdes GF, Pinheiro LC, Santos-Filho OA, Peçanha MD, Sacramento CQ, Machado V, et al. Design, synthesis, and antiviral activity of new 1H-1,2,3-triazole nucleoside ribavirin analogs. Medicinal Chemistry Research. 2014;**23**:1501-1511. DOI: 10.1007/s00044-013-0762-6

[56] Krajczyk A, Kulinska K, Kulinski T, Hurst BL, Day CW, Smee DF, et al. Antivirally active ribavirin analogues–4, 5-disubstituted 1,2,3-triazole nucleosides: Biological evaluation against certain respiratory viruses and computational modelling. Antiviral Chemistry and Chemotherapy. 2014;**23**: 161-171. DOI: 10.3851/IMP2564

[57] Cheng H, Wan J, Lin MI, Liu Y, Lu X, Liu J, et al. Design, synthesis, and in vitro biological evaluation of 1 H-1,2,3-triazole-4-carboxamide derivatives as new anti-influenza A agents targeting virus nucleoprotein. Journal of Medicinal Chemistry. 2012; **55**:2144-2153. DOI: 10.1021/jm2013503

Section 4

# Miscellaneous Applications of Azoles

Chapter 5

# Azole-Based Compounds as Corrosion Inhibitors for Metallic Materials

*Brahim El Ibrahimi and Lei Guo*


## Abstract

To face against metallic corrosion and its corresponding undesirable consequences, the implementation of corrosion inhibitor compounds is a well-known method. In this regard, a wide range of organic heterocyclic molecules has been employed as anti-corrosion agents for several metal/medium systems. Azole-based compounds, namely, N-azole, N&S-azole (i.e., thiazole), and N and O-azole (i.e., oxazole) molecules, as well as their derivatives, have shown an excellent ability to act as efficient corrosion inhibitors for different metals and alloys in various corrosive media. For this purpose, we aim in the current chapter to discuss the application of these compounds as retarders of metallic corrosion as well as related highlighted outcomes in recent years.


**Keywords:** azole, oxazole, thiazole, heterocycle, corrosion, metal, inhibitor, organic, electrochemical, surface

## 1. Introduction

Corrosion is an undesirable natural (i.e., spontaneous) phenomenon that involves the degradation of material via its electrochemical and/or chemical reactions with the components of the adjacent aggressive environment. Metals and their alloys are known as the most susceptible materials for corrosion phenomena, which are the subject of the current chapter. This spontaneous process results in significant economic and safety losses in many industrial fields, as well as in non-industrial ones [1]. According to the recent NACE's study [2], the financial loss due to the corrosion is around 2.5 trillion $ (USD), which is about 4.2% of the total gross domestic product. In the aim to face against metallic corrosion and corresponding outcomes, the implementation of corrosion inhibitor compounds is a well-known method due to its economic rentability, high efficiency, and simple utilization. By definition, a corrosion inhibitor is a chemical compound that is added, at a lower concentration, into the aggressive medium to prevent or retard (to an acceptable level) the corrosion of considered metallic material [3, 4].

There is a broad agreement in the corrosion literature that the inhibitor compounds protect metal against corrosion via their adsorption, namely, through chemical or/and physical adsorptions process, into the metal surface, which forms protective film upon the surface. Afterward, the formed compact film acts as a protective barrier on metal against aggressive species existing in the surrounding environment [5, 6]. Chemical adsorption involves the sharing of electrons between inhibitor molecules and the atoms of metal surface that leads to form coordination bonds, whereas physical bonding involves the electrostatic and/or *van der Waals* interactions between the inhibitor molecules and metal surface [7].

Among employed corrosion inhibitors in the industrial area, organic inhibitors are the most used ones, which are employed mainly in acidic media during the acid pickling, acid descaling, and acid cleaning processes of metallic materials [8, 9]. These organic compounds are characterized by the presence of lone pair electrons of heteroatoms (i.e., O, N, P, and S), functional groups (e.g., alcohols, acids, and amines), and/or multiple bonds on their molecular skeletons, which act as the favorable sites of adsorption during the inhibitor-metal interactions [10]. The adsorption process of inhibitors, hence their protection ability, is related to many factors like chemical composition and charge nature of the metal surface, electronic and molecular structures of considered inhibitor, solution's pH, temperature, inhibited solution/metal contact time, hydrodynamic conditions, and so on [11].

A wide range of organic heterocyclic molecules have been used as anti-corrosion compounds for many metal/medium systems, and others are still being explored by several researchers over the world. Especially, heterocyclic molecules containing nitrogen, oxygen, and/or sulfur atoms, such as azole, oxazole, and thiazole compounds or their derivatives, have shown remarkable protection effectiveness against metallic corrosion in several aggressive media. **Figure 1** shows the molecular structure of azole moieties, which are used as corrosion inhibitors for various metallic materials. These compounds are five-atom aromatic ring molecules that contain a nitrogen atom and at least one other nitrogen, oxygen, or sulfur atom as part of the ring [12]. The azole-based compounds can be divided into three major classes, namely, N-, N&O-, and N&S-containing azole sets. In addition to their attractive molecular structures, i.e., presence of heteroatoms, double bonds, and their planar structure, azole-based compounds are soluble in almost any polar aggressive environments, particularly in acidic media.

In this context, the inhibition of metallic corrosion by using these compounds is a well-studied academic and industrial topic. **Figure 2(a)** illustrates the number of produced publications over this topic in the last 50 years. As can be seen from this histogram, the increase of publication number demonstrates an exponential behavior, which reveals that the current topic is an active one. According to available corrosion literature (**Figure 2(b)**), nitrogen-azole derivatives (N-azoles) are extensively studied and reported as corrosion inhibitors in comparison with thiazole (N&S-azoles) and oxazole (N&O-azoles) ones. It is important to outline that recently considerable attention is directed toward the synthesizing of new azole,

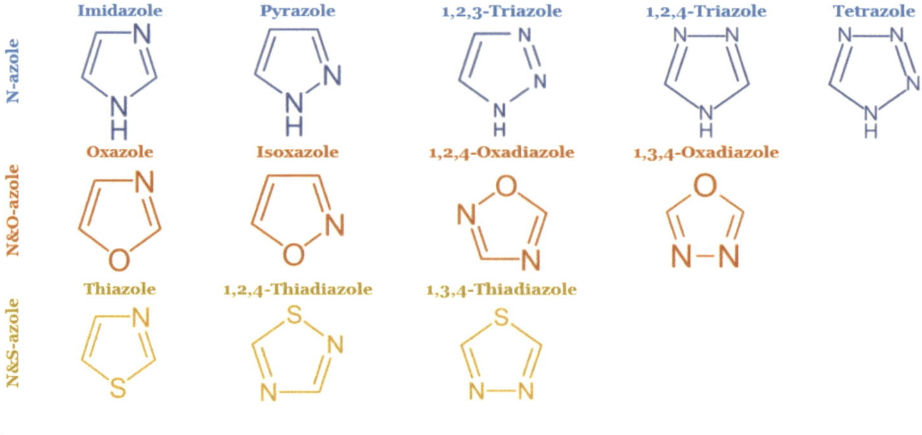


**Figure 1.**
*Molecular structures of the core rings of some azole-based compounds used as corrosion inhibitors.*

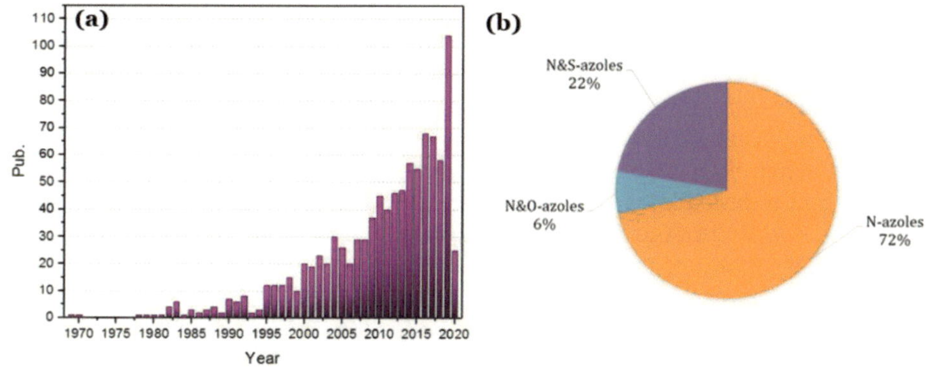


**Figure 2.**
*(a) The number of produced publications per each year from 1969 to 2020 and (b) its corresponding percentage repartition for each azole-based compounds set (i.e., N-azole, N&O-azole, and N&S-azole derivatives) according to the Scopus® database.*

thiazole, and oxazole substituted derivatives with higher prevention capacities and stability for different metal/medium combinations.

To quantify the prevention ability (i.e., inhibition efficiency) and/or to characterize the inhibition behavior/mechanism of azole-based compounds toward metallic corrosion, direct and indirect experimental techniques are used. Regarding direct techniques, they include weight loss (WL), the volume of liberated hydrogen gas (VG), and temperature variations (TV) [13–16]. Among them, the WL method is widely used because it can be employed in either concentrated or diluted corrosive solutions contrary to VG or TV ones. Besides, the indirect techniques include some direct current (DC) and alternating current (AC) electrochemical techniques, especially potentiodynamic polarization (PDP), electrochemical impedance spectroscopy (EIS), and electrochemical frequency modulation (EFM). In recent works, several researchers have limited their experimental investigations in the use of electrochemical techniques due to their high precision, the possibility to understand the action mechanism, minimal time, and material consumptions [17–19]. The inhibition efficiency of an exanimated inhibitor compounds can be calculated using Eq. ((1) in which $v^0$ and $v$ denote the corrosion rate of considered metal without and with the addition of inhibitor compound, respectively.

$$\text{IE}\ (\%) = \left(\frac{v^0 - v}{v^0}\right) \times 100 \qquad (1)$$

In the present chapter, we aim to present the application of azole-based compounds as anti-corrosion agents for metals and their alloys in the corrosive aqueous media, as well as related highlighted outcomes in recent years. For this purpose, the current chapter will be divided into three sections. We begin by the application of N-azoles as corrosion inhibitors. Afterward, we move to illustrate the main findings in the case of N&S-azoles (i.e., thiazole derivatives). Finally, we end the present chapter by their N&O-azoles (i.e., oxazole derivatives).

## 2. Using N-containing azole compounds as corrosion inhibitors for metallic materials

Among available suggestions for metal inhibition against its corrosion, N-azole compounds have shown a remarkable ability to prevent metallic degradation in different corrosive environments. For example, good inhibition effectiveness was outlined in the case of iron and copper as well as their alloys in almost any mineral acid, saline, and alkaline solutions. In this context, numerous corrosion inhibitors containing different N-azole nucleus structures (**Figure 1**) were tested and reported in the literature [20–24]. **Figure 3** displays the produced publications dealing with the inhibition of metal corrosion using these inhibitors in the latest 50 years. It is evident from this chart that triazole-based compounds are widely served as inhibitors compared to imidazole, pyrazole, and tetrazole ones, respectively. Subsequently, we will discuss the property of triazole- and imidazole-based compounds to retard the corrosion of metallic materials.

Triazole moiety can be found in numerous compounds that are used in a wide application range, especially in the medical field as antimicrobial, anti-inflammatory, anticancer, and antifungal drugs [20, 21]. On the other hand, the existence of three nitrogen atoms in the same molecule with a planar geometry has attracted the attention of many corrosion scientists to evaluate the protective effect of triazole molecules against metallic corrosion. Good inhibition property of either 1,2,4- or 1,2,3-triazole molecules is noted in the case of various metals, e.g., copper, iron, and its alloys in acid and non-acid media [22–24]. Recently, more attention has been focused on the development of new stable anti-corrosion compounds containing triazole core rings [25, 26]. As a result, these compounds have shown a remarkable affinity toward metallic surfaces, leading to the formation of a protective organic film on the surface of the protected metal. Furthermore, in most cases, the inhibition efficiency of these compounds increases by increasing their concentration. The role of triazole-based compounds as corrosion inhibitors for copper, iron, aluminum, zinc, and its alloys has been outlined in many corrosive media [27–31]. Among considered media, there are $H_2SO_4$, HCl, $HNO_3$, $H_3PO_4$, and NaOH solutions at different concentrations, as well as natural/artificial seawater, sulfate, and chloride environments [32]. **Figure 4** shows the molecular structures of some 1,2,4-triazole derivatives used as effective corrosion inhibitors.

It was found that the nature of side substitutions of the triazole moiety has strongly influenced its ability to prevent corrosion phenomena. For instance, Resende et al. [33] have evaluated the inhibition capacity of three newly synthesized 1,2,3-triazole derivatives (**C-1**, **C-2**, and **C-3**, **Figure 5**) through click chemistry reaction against carbon steel corrosion in acid media. They observed that the recorded inhibition efficiency of these heterocyclic molecules depends on the substituent nature, which is ranked as **C-2** (96%) > **C-1** (92%) > **C-3** (72%) at 250 mg $L^{-1}$ of inhibitors after 24 hours of immersion. Moreover, **C-2** and **C-1** inhibitors exhibited an excellent inhibition trend in comparison with a commercial inhibitor as reported by the authors. In another study, additionally to O&N

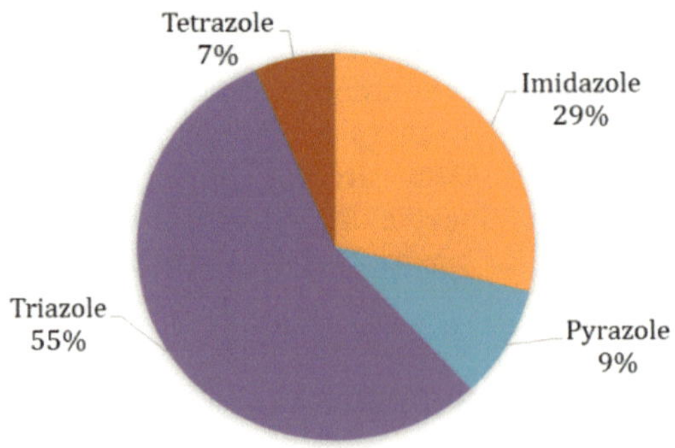


**Figure 3.**
*Distribution of produced publications related to the use of N-azole family corrosion inhibitors according to the Scopus® database.*

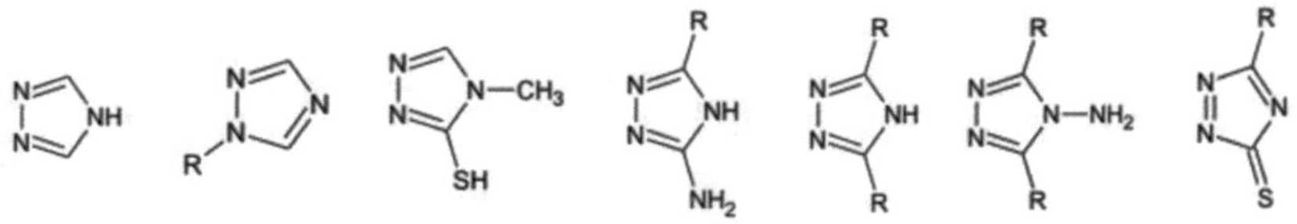

**Figure 4.**
*Some 1,2,4-triazole-based compounds used as corrosion inhibitors.*

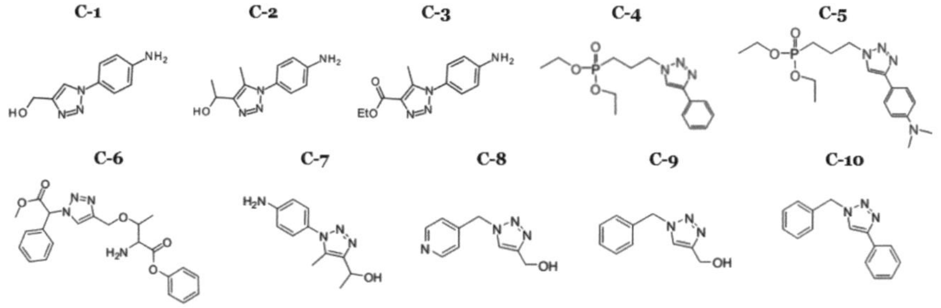


**Figure 5.**
*Some 1,2,3-triazole derivatives used as corrosion inhibitors.*

heteroatoms and phenyl rings characterizing **C-1** and **C-2** compounds, the introduction of phosphorus atom (P) was done to synthesis two new ecologically 1,2,3-triazole derivatives (**C-4** and **C-5**, **Figure 5**). The corrosion assays demonstrated that the addition of dimethylamino (—$N(CH_3)_2$) functional group in the side phenyl ring has improved the prevention efficiency of newly examined inhibitors from 91 to 94% at 1 mM for mild steel in 1 M HCl solution. An inhibition efficiency over 80% is also achieved by using other 1,2,3-triazole derivatives, e.g., **C-6**, **C-7**, **C-8**, **C-9**, and **C-10** in **Figure 5** [34–37].

It is well-known for more than 60 years that the combination of triazole core ring with benzene one, the so-called benzotriazole (**C-1** in **Figure 6**), as well as their derivatives can act as efficient and stable corrosion inhibitors during long contact time for several metal/solution systems, especially for copper and its alloys [38]. For instance, this bicyclic aromatic molecule behaves as a useful inhibitor for pure copper, $Cu_{90}Zn_{10}$, and $Cu_{60}Zn_{40}$ alloys in chloride environments such as 3.5% NaCl solutions and artificial seawater [39, 40]. The good corrosion prevention capacity was also obtained both for dynamic and stagnate conditions at lower concentrations. Nonetheless, lesser inhibition efficiencies of benzotriazole and its derivatives are gained in acidic media than the base and near-neutral ones, which is due to the dissolution of formed protective film on the metal surface in acid media [38, 41]. A literature examination discloses that benzotriazole showed a particular ability to control the corrosion of AA2024 aluminum alloy in 5 mM NaCl solution as compared to 1,2,4-triazole and amino-1,2,4-triazole, and in its presence both anodic and cathodic dissolutions were reduced [27]. Additionally, in sulfide-polluted 3.5% NaCl solution, an excellent inhibition performance of 93% is obtained for carbon steel at 5 mM of benzotriazole [42].

As the main way to enhance the capability of benzotriazole to control metallic corrosion, there is the chemical modification of its molecular structure. This strategy aims to introduce more adsorption sites within the benzotriazole skeleton by adding functional groups and conjugated systems. In this regard, various benzotriazole-based derivatives were synthesized and tested as corrosion inhibitors. For instance, a new heterocyclic derivative consisting of two benzotriazole molecules and 1,3,4-thiadiazole moiety (**C-2**, **Figure 6**) exhibited good inhibition efficiency for copper in chloride environments both at acidic and near-neutral pH, 79

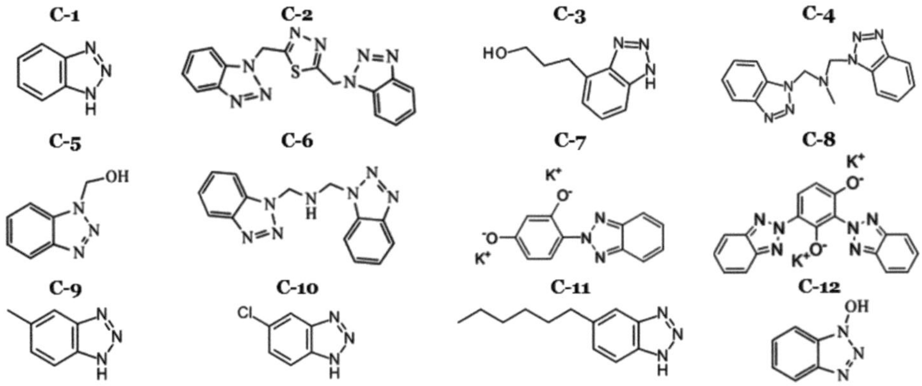


**Figure 6.**
*Some benzotriazole family corrosion inhibitors.*

and 87% at 1 mM, respectively [41]. Recently, two structural benzotriazole derivatives (**C-3** and **C-4**, **Figure 6**) have been reported as useful anti-corrosion compounds against the degradation of brass alloy in an artificial seawater. For example, at 150 ppm the inhibitors offer 82 and 92% as corrosion reduction percentages for **C-4** and **C-3**, respectively [39]. Furthermore, Ravichandran et al. [43] have carried out a comparative study on three benzotriazole-based inhibitors, namely, **C-1**, **C-5**, and **C-6** as depicted in **Figure 6**, for brass alloy corrosion in 3% NaCl solution. The associated outcomes of this study reveal that all tested heterocyclic molecules behave as efficient corrosion inhibitors and the inhibition efficiency increases as follows: **C-1** (77%) < **C-5** (90%) < **C-6** (93%) at lower concentration (150 ppm). The observed protection is attributed to the formation of inhibitor Cu(I) complexes on the metal surface, which isolate the surface from aggressive agents in the solution. Many other novel benzotriazole derivatives with more or less complex molecular structures have been reported in the literature as potent anti-corrosion compounds such as **C-7**, **C-8**, **C-9**, **C-10**, and **C-11** derivatives in **Figure 6** [44–46].

It is important to specify that the introduction of further functional groups into the benzotriazole skeleton has not usually improved its inhibition performance. For instance, it was outlined that the alcohol-benzotriazole derivative (**C-12**, **Figure 6**) exhibited reduced inhibition efficiency compared to simple benzotriazole for pure copper immersed in 3% NaCl medium [47]. Besides, without performed additional chemical modifications on the benzotriazole molecular skeleton, the improvement of its inhibition performance can be also done via the synergism effect with other additive species, e.g., halide and metallic ions and organic and inorganic compounds [48, 49]. As reported by Bokati et al. [50], the addition of phosphate ($Na_3PO_4$) and molybdate ($Na_2MoO_4$) compounds into corrosive solution (natural seawater) have enhanced the inhibition efficiency of benzotriazole, particularly for copper, as compared to mild steel alloy. Additionally, the mixture of benzotriazole/$Ce^{3+}$ was proven to have greater synergistic inhibition effect for zinc/iron and aluminum/copper model galvanic couples in NaCl solution [51, 52].

An additional N-containing azole variety compound that has also received sufficient attention is imidazole and its derivatives as well. Such attention is due to its non-toxicity and appropriate molecular and electronic structures to act as a corrosion inhibitor: the compound is planar and aromatic and contains 2 N heteroatoms. Its mechanism of action as an inhibitor is the same as stated for other reported azole compounds. An increase of concentration leads to an enhancement of its protection capacity, while in many cases the temperature has shown an undesirable effect: its increase can imply a reduction of observed inhibition property of imidazole-based inhibitors. The tendency of imidazole heterocyclic molecules to inhibit metal corrosion, especially for copper, has been extended to synthesis novel derivatives having excellent inhibition efficiency for a longer time. The latter extension aimed to introduce additional favorable centers of adsorption via some functional or non-functional groups such as —SH, —$NH_2$, —COH, —$OCH_3$, —$SCH_2$Phe, and —Phe [53–57]. **Figure** 7 illustrates the chemical structure of some substituted imidazole moieties used as corrosion inhibitors. It was outlined that imidazole-based compounds showed interesting activity to act as anti-corrosion agents in several corrosive environments like $HNO_3$, HCl, $H_2SO_4$, NaCl, and NaOH media, with the higher prevention efficiencies noted in chloride and in sulfuric acid solutions. **Table 1** shows the inhibition data related to the application of some imidazole derivative (**Figure** 7) as retarder compounds against copper corrosion in various media [58–64]. On the other hand, the synergism effect has also been used to improve further the attained inhibition efficiency, which was performed by adding supplementary additives, e.g., halide ions, into the inhibited solution [60].

Another common anti-corrosion compound among imidazole-based derivatives is benzimidazole, which is a heterocyclic aromatic molecule with planar geometry consisting of an imidazole and a benzene moiety (**C-1**, **Figure 8**). It was discovered for the first time by Hoebrecker as a part of vitamin $B_{12}$ [65]. In the last decades, benzimidazole, as well as its derivatives, has been reported as effective anti-corrosion agents for many metallic materials such as mild and carbon steels [66, 67]. The property of benzimidazole-based inhibitor to retard corrosion rate was attributed to the formation of an adsorbed protective film on the metal surface, which can consist of metal-benzimidazole complex or adsorbed benzimidazole molecules [68, 69]. As stated for benzotriazole, numerous benzimidazole derivatives with different structural compositions have been synthesized and then used as corrosion inhibitors. In this regard, simple benzimidazole derivatives showed potent inhibition effect, and in order to obtain them the chemical modification of benzimidazole core is carried out by the insertion of different functional groups. Among introduced groups, there are —SH, —$NH_2$, —OH, —$SCH_3$, —$CH_2NH_2$, —$CH_2OH$, —Cl, —Br, and carbon chain with different lengths [70–78]. **Figure 8** summarizes the chemical structures of some benzimidazole-based derivatives employed as corrosion inhibitors and their corresponding inhibition data.

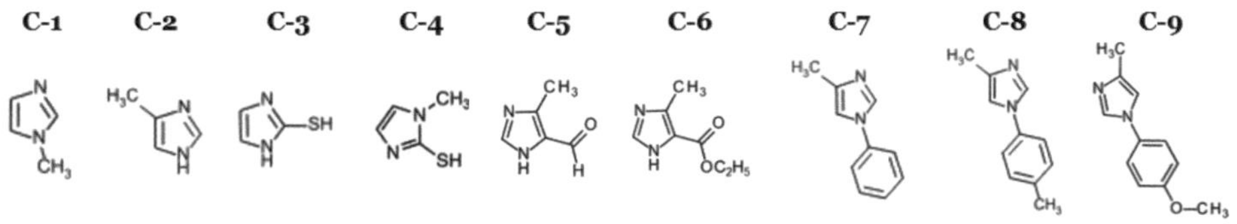


**Figure 7.**
*Molecular structures of some substituted imidazole derivatives used as corrosion inhibitors.*

| Inhibitor | Media | IE ([inh.]) |
|---|---|---|
| Imidazole | 0.5 M $H_2SO_4$/3% NaCl/0.1 M NaOH | 55% (0.5 M)/50% (0.1 mM)/46% (2 mM) |
| C-1 | 1 M $H_2SO_4$/1 M HCl/3% NaCl | 70% (10 mM)/90% (10 mM)/61% (10 mM) |
| C-7 | 0.5 M $H_2SO_4$/3% NaCl | 93% (0.5 M)/94% (5 mM) |
| C-8 | 0.5 M $2SO_4$/0.5 M HCl/3%NaCl | 88% (0.05 M)/54% (0.1 M)/93% (0.7 mM) |

*[inh.]: inhibitor concentration.*

**Table 1.**
*The inhibition efficiency (IE) of imidazole and some of its derivatives (see **Figure 7**) against copper corrosion.*

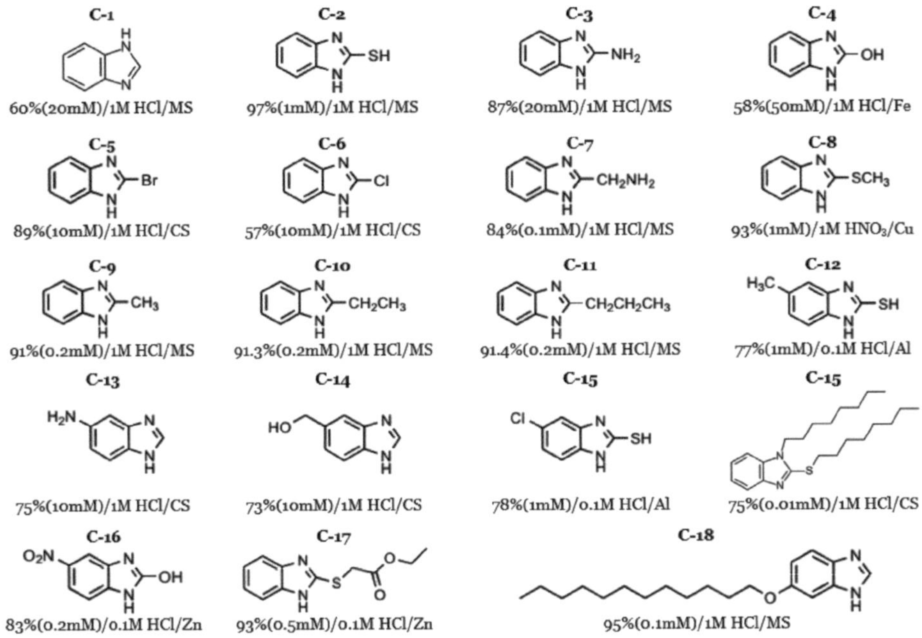


**Figure 8.**
*Molecular structures of some reported benzimidazole-based derivatives as corrosion inhibitors, as well as corresponding inhibition data, which are presented as "inhibition efficiency, % (inhibitor concentration, mM)/ corrosive medium, M/metal." Abbreviations: CS, carbon steel; MS, mild steel; Fe, pure iron; Cu, copper; Zn, zinc; Al, aluminum.*

An additional strategy to enhance the performance of benzimidazole to inhibit metallic corrosion is the combination of the latter heterocyclic molecule with other aromatic systems like benzene or triazole core rings without and with further substituent groups. Under this view, various hybrid benzimidazole/aromatic ring-based derivatives have been reported as anti-corrosion molecules [79–86]. **Figure 9** summarizes some benzimidazole/aromatic ring class inhibitors, as well as corresponding inhibition data. Even in very corroding media, benzimidazole/aromatic ring derivatives have shown excellent ability to protect metallic materials against corrosion in these media. For instance, it was found that **C-9** and **C-10** (**Figure 9**) derivatives could offer good protection against mild steel corrosion in a 15% HCl solution. The maximum corrosion retardation of 91% was pointed out for **C-9** derivative with —$OCH_3$ side phenyl substituent at 200 ppm concentration [87, 88].

On the other hand, several simple and complex bridged benzimidazole derivatives (i.e., bis-benzimidazoles) were employed as potent corrosion inhibitors in which different chain bridges are implemented as linear carbon chains without and with heteroatoms. **Figure 10** presents some bis-benzimidazole corrosion retarders. For instance, 1,4-bis-benzimidazolyl-butane (**C-1** in **Figure 10**) exhibited an efficiency of 98% at 0.68 mM inhibitor for mild steel in acid media [89], while at lower concentration (0.10 mM) the insertion of a nitrogen atom in the carbon bridge (**C-2, Figure 10**) provided good inhibition efficiency of 89% [71]. Ahamad et al. [90] reported the connection of two benzimidazoles via di-sulfur-bridge for the synthesis of the novel derivative (**C-3, Figure 10**). The corrosion tests reveal the excellent property of bridged benzimidazole inhibitors to control mild steel corrosion both in hydrochloric and in sulfuric acid media, with the attained inhibition efficiencies around 98%. Furthermore, it was found that some bis-benzimidazole derivatives can offer higher inhibition prevention for prolonged immersion time as

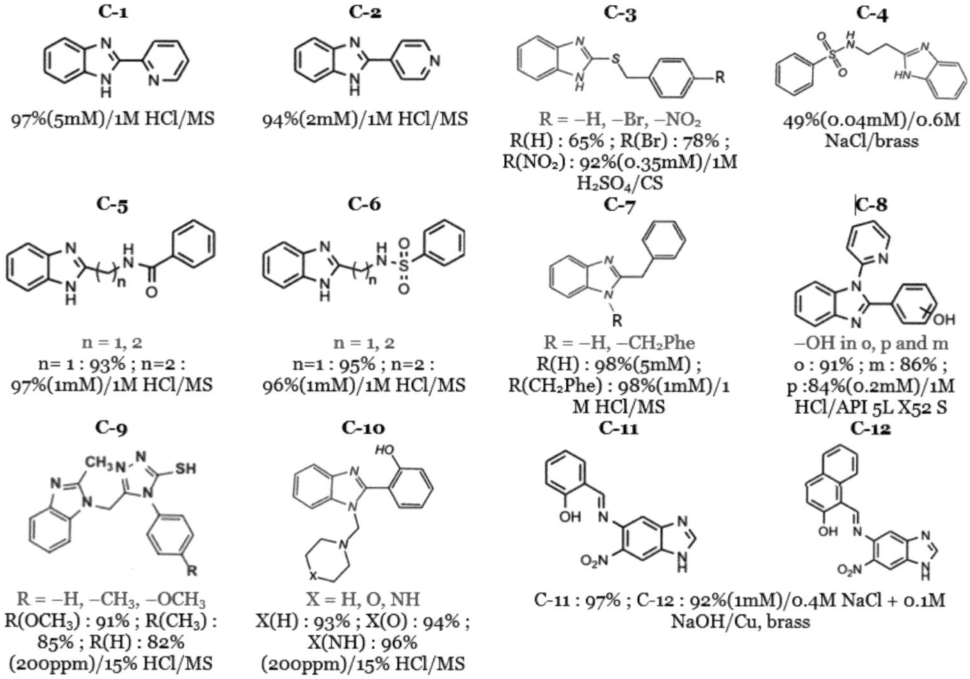


**Figure 9.**
*Molecular structures of some reported benzimidazole/aromatic ring derivatives set as corrosion inhibitors as well as corresponding inhibition data.*

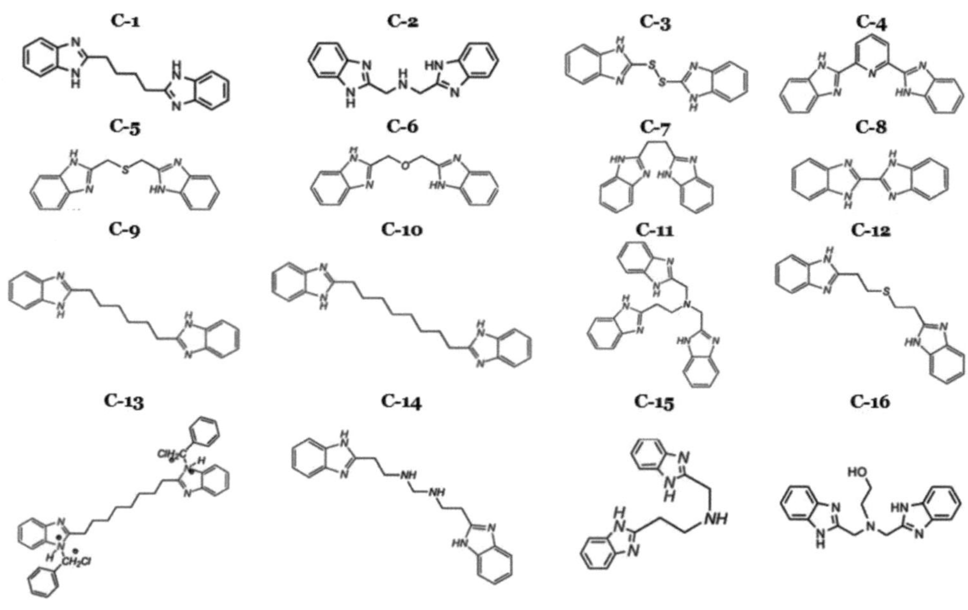


**Figure 10.**
*Molecular structures of some used bis-benzimidazole corrosion retarder's type.*

reported by Dutta et al. [91] for **C-4**, **C-5**, **C-6**, and **C-7** compounds, with the lower recorded efficiency 88% after 4-day immersion of mild steel in 1 M HCl solution. The length of the carbon chain of the benzimidazole bridge has influenced the ability of these derivatives to retard corrosion. In this context, three bis-benzimidazole derivatives (**C-8**, **C-9** and **C-10**, **Figure 10**) exhibited a significant tendency to reduce mild steel corrosion in acid environment, with an inhibition percentage up to 94% obtained at 0.1 mM for the derivative with longer carbon chain (i.e., **C-10**).

To understand the action mechanism of an inhibitor compound at an atomic scale, the calculation of some electronic and molecular parameters using a chemical computational approach corresponding to the adsorption process is required. In this context, Kokalj's team and other groups have studied in-depth the role of the molecular and electronic structures of many N-azole inhibitor molecules for their inhibition property for various metallic materials [92–97]. Density functional theory (DFT)-based calculations have been employed by these scientists to quantify the interaction magnitude of considered inhibitor molecules with the chosen metal surfaces, as well as their adsorption configuration onto these surfaces through qualitative analysis.

## 3. Using N&S-containing azole compounds (thiazoles) as corrosion inhibitors for metallic materials

Referring to previous works [98–100], heterocycle-based inhibitors with both sulfur and nitrogen atoms in their structure were offered outstanding prevention activities in comparison with those containing only sulfur or nitrogen atoms. In this regard, several N&S-containing azole compounds (**Figure 1**), like thiazole and thiadiazole derivatives, have been attested to be operational inhibitors against the corrosion of many metallic materials in a wide variety of corrosive media. Based on the available corrosion literature, special attention is devoted to thiadiazole-based compounds compared to thiazole ones. Such attention trend is based on the fact that the presence of further heteroatoms (N atoms) on those heterocyclic molecules can raise their adsorption onto the metal surface and consequently enhance their inhibition effectiveness.

In addition to the potent affinity of pre-existing heteroatoms (i.e., N and S atoms) in the 1,3-thiazole ring to interact with the metal surface during the inhibition process, the attachment of the latest ring with further substituents to improve its inhibition efficiency was recently reported. In this view, many 1,3-thiazole-based derivatives are developed via different synthesizing reaction procedures. Conferring to obtained results, these new derivatives were shown to have a great tendency to reduce the dissolution of various metallic substrates. For instance, Raviprabha and Bhat [101] have evaluated the anti-corrosion property of ethyl-2-amino-4-methyl-1,3-thiazole-5-carboxylate derivative (**C-1, Figure 11**) for AA6061 aluminum alloy in 0.05 M HCl medium. Based on the calculated thermodynamic parameters corresponding to the adsorption process of **C-1** molecules, the chemisorption process of derivative molecules is proposed as a potential mechanism of inhibition. Moreover, it was disclosed that an increase in temperature level implies an elevation of inhibition activity of evaluated 1,3-thiazole derivative, with the prevention percentage of 93% at 333 K and 100 ppm of **C-1**. Another similar 1,3-thiazole derivative (**C-2, Figure 11**) with pyridinium ring also showed a good capacity to regulate copper dissolution in molar HCl solution, with a maximum of 94% as prevention efficiency achieved at $10^{-3}$ M.

The nature and position of added substituents in a 1,3-thiazole ring-based inhibitor can considerably influence its inhibition performance. Recently, two mono-substituted 1,3-thiazole derivatives (**C-3** and **C-4, Figure 11**) have revealed this behavior, which were used to protect X65 steel alloy largely employed in pipelines for natural gas transportation purposes. The ethenone-substituted 1,3-thiazole derivative (**C-4**) exhibited superior performance to control X65 steel dissolution than isobutyl one (**C-3**), in which the recorded prevention efficiency at $5 \times 10^{-3}$ M is being around 90 and 70%, for **C-3** and **C-4**, respectively [102].

In addition to lateral substituents, which contain supplementary electron-donating centers (e.g., functional groups, aromatic and azole rings), the inhibition performance of 1,3-thiazole-based derivatives is also improved by increasing their electron-donating capability via attachment with a benzene ring. In this regard, Chugh et al. [103] have synthesized four new derivatives based on benzo[d]thiazole core structure (**C-5**, **Figure 11**), which exhibited an increased anti-corrosion property by replacing hydrogen atom (IE = 79%) of R substituent (on the lateral benzene ring) by chlorine atom (IE = 85%), methyl group (IE = 88%), and finally —$NH_2$ functional group (IE = 90%). In the same way, the combination of benzo[d]thiazole bi-rings with imidazoline ring (**C-6**, **Figure 11**) is found to act as an efficient corrosion inhibitor in the water-glycol medium [104]. More complex 1,3-thiazole derivative molecules were evaluated and reported as good corrosion inhibitors at lower concentrations, e.g., ceftriaxone 1,3-thiazole derivative (**C-7**, **Figure 11**) demonstrated an inhibition percentage of 95% at 400 ppm for mild steel in acidic environment [105]. **Table 2** illustrates the relevant outcomes on the use of two other 1,3-thiazole-based compounds as corrosion inhibitors [106, 107].

1,3,4-Thiadiazoles, another class of thiazole heterocyclic molecules, have been widely examined for their uses in numerous fields such as agrochemical and pharmaceutical areas. For example, sulfamethoxazole and methazolamide are market drugs that contain a 1,3,4-thiadiazole ring [108, 109]. On the other hand, the use of 1,3,4-thiadiazole-based compounds as inhibitor additives also reduced the degradation of metals caused by the surrounding aggressive environment. Many 1,3,4-thiadiazole derivatives were reported to act as potent anti-corrosion agents in different operating conditions. The molecular structure of this five-atom ring type is characterized by the incorporation of an additional nitrogen atom into the 1,3-thiazole ring in 4 position. The presence of further heteroatoms in conjugated 1,3,4-thiadiazole-based molecules plays a curious role in their protection activities. The latest feature is due to the highest tendency of heteroatoms with the conjoint multi-bonds to facilitate the adsorption of these compounds onto the metal surface, and subsequently formed protective film isolates the substrate from solution components.

Several 1,3,4-thiadiazole derivatives with different attached hydrocarbon chains were synthesized and evaluated as corrosion inhibitors. It was found that the size and shape of inserted substituents, as well as their chemical properties, can influence the performance of developed 1,3,4-thiadiazole derivatives to retard metal dissolution. For instance, the substitution of mercapto groups at 2 and 5 positions of the thiadiazole nucleus by ethyldisulfanyl (**C-1**, **Figure 12**) augmented the achieved

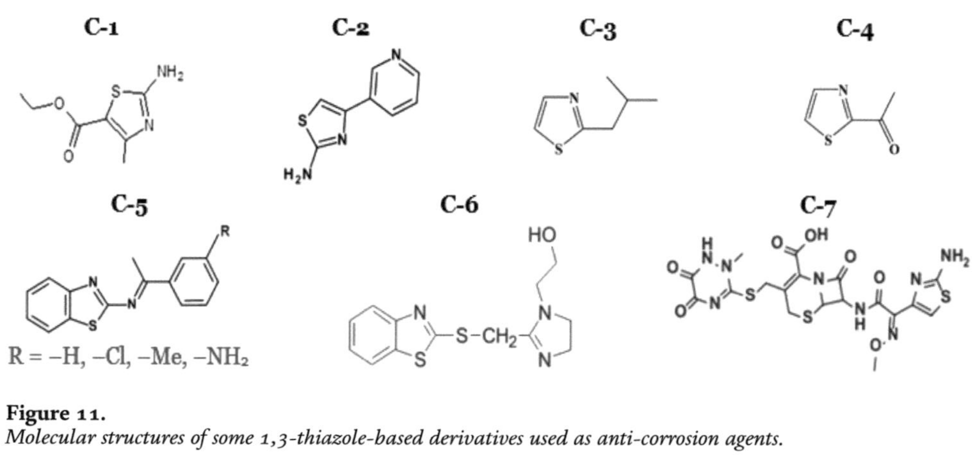


**Figure 11.**
*Molecular structures of some 1,3-thiazole-based derivatives used as anti-corrosion agents.*

inhibition efficiency from 82.4 to 88.1% at 0.4 mM of inhibitors toward copper corrosion in PAO base oil environment. Concerning the protection activity of these compounds, it was attributed to their physical adsorption on copper oxide surface as theoretically expected and experimentally verified [110]. Moreover, a series of 2,5-dimercapto-1,3,4-thiadiaxole derivatives was also reported as anti-corrosion compounds by Wei and Gemmill et al. [111, 112]. Molecular structures of some reported derivatives are summarized in **Figure 13**.

In recent years, microwave irradiation heating has been used as a convenient green method for the synthesis of different heterocyclic inhibitors [113] from which we mention 2-amino-5-alkyl-1,3,4-thiadiazole derivatives, with the corresponding synthesizing scheme displayed in **Figure 14**. The length of the side alkyl chain impacted their capacity to control the dissolution of mild steel in 1 M $H_2SO_4$ solution, with the inhibition effectiveness increase with rising chain length, except for —$C_{13}H_{27}$ alkyl case for which the prevention efficiency rapidly decreases [114]. Additionally, the replacement of the alkyl chain of 2-amino-5-alkyl-1,3,4-thiadiazole by mercapto substituent (—SH) was led to a perfect protection efficiency of 99.3% [115].

On the other hand, four novel 1,3,4-thiadiazole-thiosemicarbazones derivatives and their cobalt(II) ion complexes (**C-2, Figure 12**) have been found to play the important role as anti-corrosion agents for carbon steel in acid media. However, the tests revealed that the molecular structure of these compounds has a little effect on the obtained inhibition efficiencies, which are around 90% in the presence of 500 ppm inhibitors [17]. Based on 1,3,4-thiadiazol-2-amine, new heterocyclic scaffold derivative (**C-3, Figure 12**) was synthesized and reported as an excellent inhibitor (IE = 91% at 0.5 mM) against mild steel corrosion in the molar hydrochloric acid medium [116, 117]. Another derivative of 1,3,4-thiadiazol-2-amine (**C-4, Figure 12**) has been also reported to act as a useful inhibitor for copper in de-aerated, aerated, and oxygenated 3% NaCl solutions, with a maximum efficiency of 94% obtained at 5 mM of the inhibitor [118]. Besides, 1,3,4-thiadiazol-containing organic inhibitors also served to improve the anti-corrosion property of

| 1,3-Thiazole derivative | Metal | Medium | Inhibition efficiency (IE) |
|---|---|---|---|
| 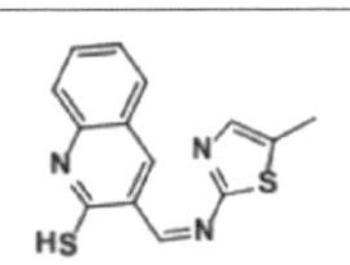 | Mild steel | 1 M HCl | **IE** = **96%** at 298 K and 5 mg $kg^{-1}$ |
| 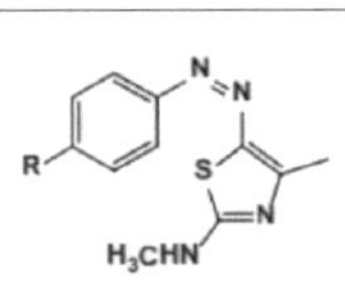 R = -H, $-CH_3$, $-OCH_3$, -Br, $-NO_2$ | 304 L stainless steel | 3 M HCl | **IE:** $OCH_3$ **(93%)** > $CH_3$ **(91%)** > H **(89%)** > Br **(87%)** > $NO_2$ **(86%)**. IE was enhanced by synergistic combination with 1 mM KSCN at 303 K and 11 × $10^{-3}$ mM |

**Table 2.**
*Relevant data related to the application of some 1,3-thiazole-based compounds as corrosion inhibitors.*

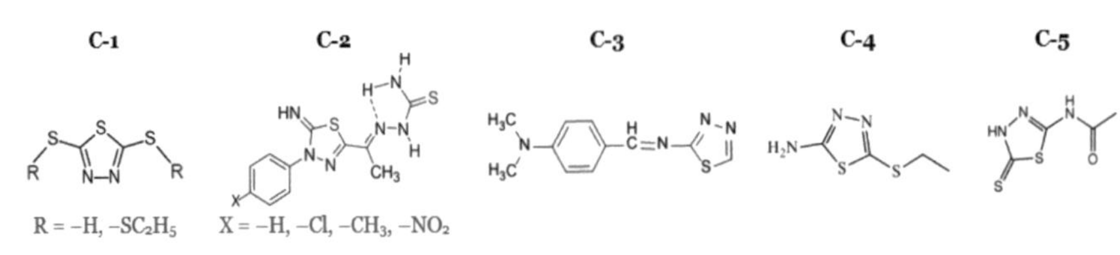


**Figure 12.**
*Molecular structures of some 1,3,4-thiadiazole-based derivatives used as anti-corrosion agents.*

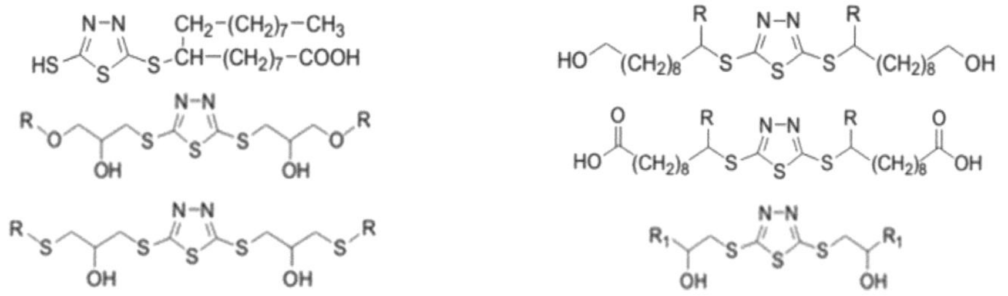

**Figure 13.**
*Some 2,5-dimercapto-1,3,4-thiadiaxole anti-corrosion compounds.*

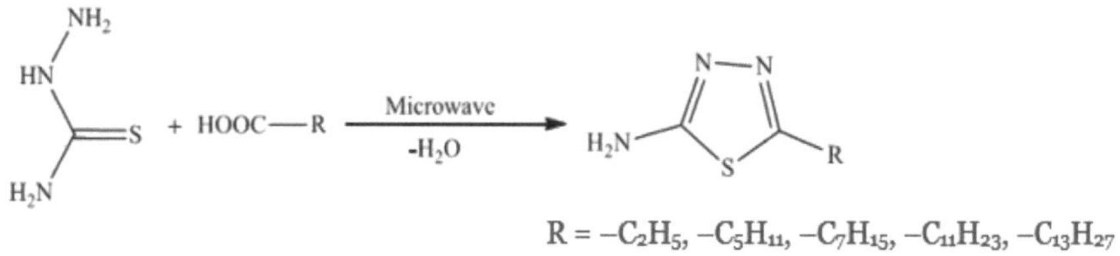

**Figure 14.**
*Synthesis route of 2-amino-5-alkyl-1,3,4-thiadiazole derivatives under microwave irradiations.*

some coatings. For instance, 2-acetylamino-5-mercapto-1,3,4-thiadiazole (**C-5, Figure 12**) has shown a good ability to improve the protective quality of chitosan coatings on zinc, for which the protection efficiency passed from 64 to 91% in the presence of **C-5** derivative [119].

## 4. Using N&O-containing azole compounds (oxazoles) as corrosion inhibitors for metallic materials

In addition to N&S-containing azole corrosion inhibitors, oxazole-based compounds (i.e., N&O-containing azoles) have gained considerable attention in recent years in this regard. Oxadiazole molecule consists of a five-membered heterocyclic ring with at least one nitrogen and an oxygen atom. These N&O-containing heterocycles are interesting molecules that exist in wide biological-based compounds like diuretics, anxiolytics, and local anesthetics. Moreover, oxazole shows an antimycotic activity and can be used as anti-inflammatory agents as well as antibacterial toward pneumoniae, micrococcus, and *Staphylococcus aureus* [120, 121].

Numerous N&O-containing azole heterocyclic molecules have been studied and reported as efficient anti-corrosion agents for various metallic materials, especially in acidic media [122–126]. Such beneficial effects are related to their special affinity to adsorb on the metallic surfaces. Moreover, these compounds possess lone pair electrons on the oxygen and nitrogen atoms, which can interact favorably with the vacant orbitals of metal, leading to formation of protective barrier film [127]. **Figure 15** shows the produced publications related to the inhibition of metal corrosion employing oxazole-based inhibitors in the last 50 years. It is clear from this figure that among available N&O-containing azole compounds, the oxazole, isoxazole, 1,2,4-oxadiazole, and 1,3,4-oxadiazole ones (**Figure 1**) are frequently used for corrosion inhibition purposes. Noticeable attention is focused on 1,2,4- and 1,3,4-oxadiazole inhibitors, mainly due to the presence of several nitrogen atoms on their five-membered heterocycle in comparison to other N&O-azoles.

Due to its excellent descaling properties, sulfamic acid ($NH_2HSO_3$) is used in a large variety of industrial applications such as cleaning of heat exchangers and cooling water systems. As compared to other acids like hydrochloric acid, sulfamic acid shows a lower corrosion rate of stainless steel (SS) without the problem of chloride-induced stress corrosion cracking of SS. In order to reduce further this corrosion, the addition of inhibitor compounds into sulfamic media is mandatory. As effective inhibitor candidates, four new synthesized oxazole derivatives have been reported as good corrosion inhibitors for 316 L-type SS in 0.6 M $NH_2HSO_3$ solution by Fouda et al. [122]. The molecular structures of reported oxazole derivatives are presented in **Figure 16** (**C-1**, **C-2**, **C-3**, and **C-4**). According to weight loss experiments and electrochemical tests via different techniques, a good prevention ability around 90% is recorded at lower concentration (i.e., $2 \times 10^{-4}$ M) of investigated derivatives after a moderate immersion time (3 h), especially for the fourth derivative (**C-4**). In addition to the presence of benzene ring and nitrogen and oxygen atoms, the good inhibition property of **C-4** derivative as compared to the other ones is attributed to the existence of four aromatic rings as substituents, which results in its larger molecular size and planar geometry, leading to highest coverage of the metal surface area by adsorbed **C-4** molecule. Based on this study, it can be outlined that the substitution of oxazole core ring by biggest lateral substituents can effectively improve the inhibition property of oxazole derivatives at lower concentrations.

The protection activity of other oxazole derivatives set has been reported in recently published work [123]. The authors of this work have synthesized a series of three 2-phenyl oxazole derivatives with different substitutions at the carbon five of the oxazole ring (**C-5**, **C-6**, and **C-7**, **Figure 16**). A significant reduction of mild steel dissolution rate in molar hydrochloric acid solution is observed in the presence of these derivatives. The protective effect of synthesized oxazole compounds can be clearly revealed in **Figure 17**, in which i-E curve decrease is shown in the presence of these compounds as their concentrations rise (colored lines) compared to the blank solution (black line). Accordingly, the order of corrosion inhibition is as follows: **C-6** (94.7%) > **C-7** (85.9%) > **C-5** (78.6%) at $10^{-3}$ M concentration. Using quantum chemical computations via the DFT-B3LYP/6-31G(d,p) method, the highest inhibition activity of the **C-6** oxazole derivative is attributed to its great reactivity with the metal surface, which is induced by the benzene-1-sulfonate substituent. The presence of sulfur atom can cause the elevation of oxazole compounds adsorption process onto the metal surface, which reflects the good prevention capacity of these compounds.

In order to get great protection performance, the synthesis of new oxazole derivatives in which other azole-based core rings are incorporated has been reported. In this context, three new benzoimidazole/1,3,4-oxadiazole derivatives (**C-1**, **C-2**, and **C-3**, **Figure 18**) were reported as efficient organic inhibitors for mild steel dissolution

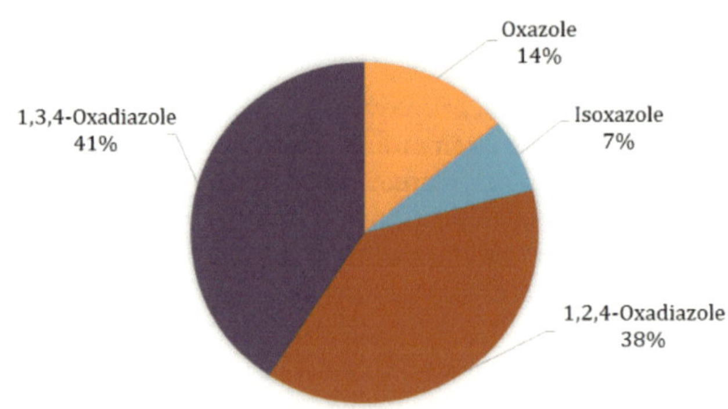


**Figure 15.**
*Distribution of produced publication percentage related to the corrosion inhibition using oxazole-based compounds according to the Scopus® database.*

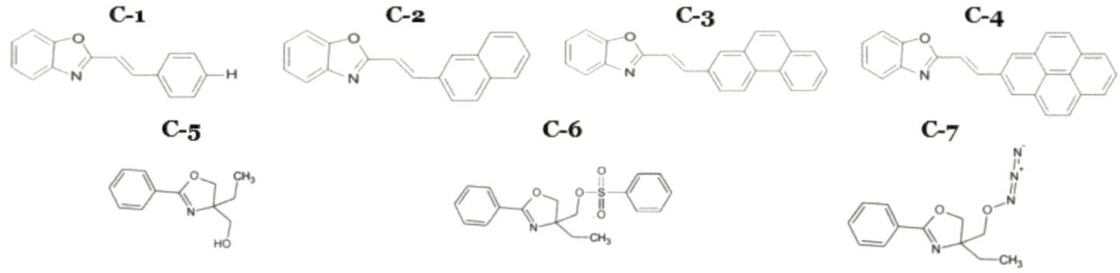


**Figure 16.**
*Chemical structures of newly synthesized benzo and 2-henyl oxazole derivatives.*

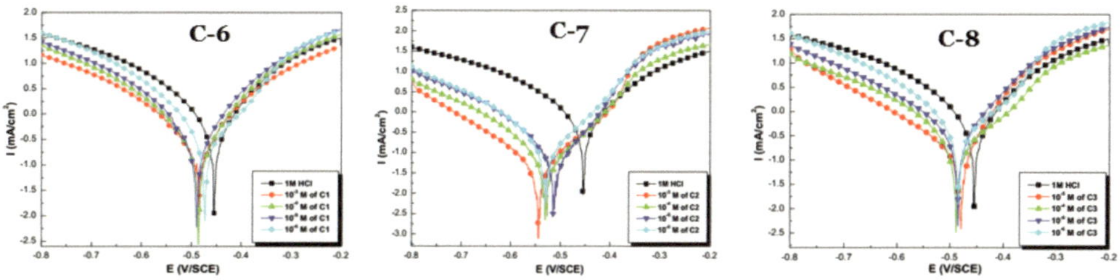


**Figure 17.**
*Potentiodynamic curves of mild steel in 1 M HCl without and with synthesized 2-phenyl oxazole derivatives (**C-6**, **C-7**, and **C-8**, **Figure 16**) at different concentrations [112].*

in acidic solutions [124–126]. These compounds exhibited an interesting effect in both sulfuric and hydrochloric acid solutions, which are largely used for the metal cleaning process in several industrial fields. It should be kept in mind that the higher reduction of corrosion rate caused by adding these inhibitors is obtained in hydrochloric acid than the sulfuric one, which reveals the possible effect of aggressive

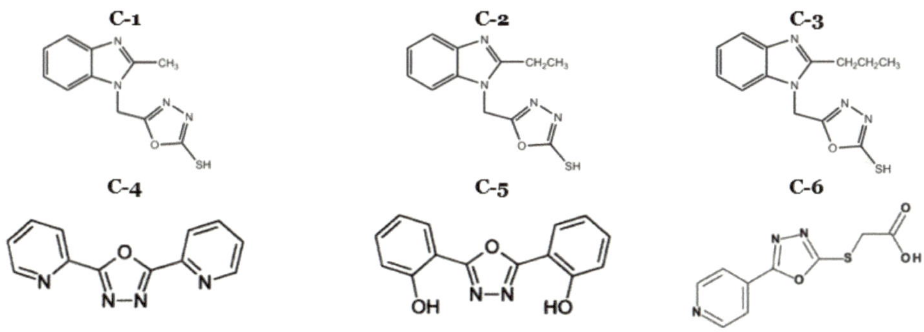


**Figure 18.**
*Chemical structures of examined 1,3,4-oxadiazole derivatives.*

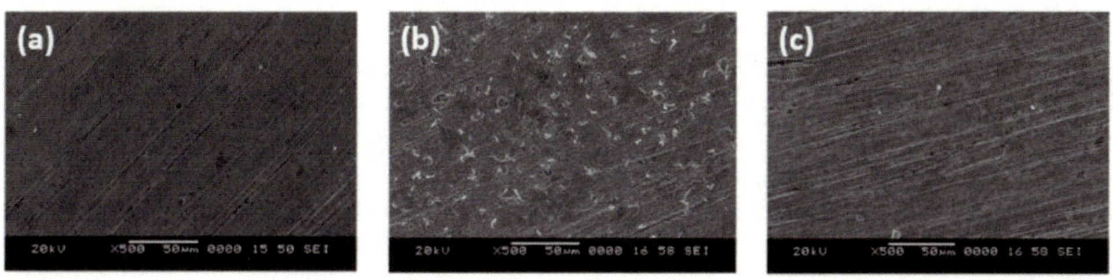


**Figure 19.**
*SEM images of mild steel samples (a) before and after immersion in 0.5 M HCl solution, (b) without and (c) with **C-2** benzoimidazole/1,3,4-oxadiazole derivative [115].*

C-1 C-2 C-3

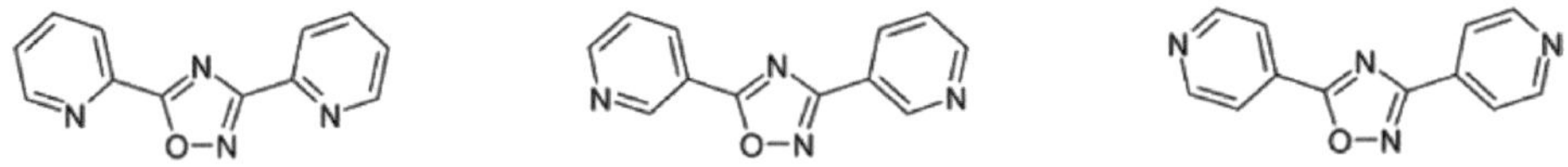

**Figure 20.**
*Chemical structures of 1,2,4-oxadiazole derivatives with pyridinium substituents.*

C-1 C-2 C-3

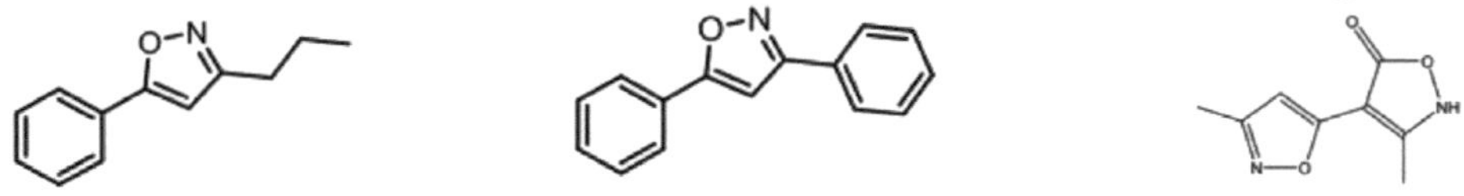

**Figure 21.**
*Chemical structures of isoxazole derivatives.*

media on the inhibition activity of used benzoimidazole/1,3,4-oxadiazole inhibitors. Moreover, the nature of considered acid can influence also the trend of recorded inhibition efficiencies, e.g., in HCl solution; the order is **C-1** (≈92%) > **C-2**, while in $H_2SO_4$ one is **C-2** (75%) > **C-1** > **C-3** at the same concentration. Such conclusions are in good agreement with those of Bentiss et al. [128], which used other 1,3,4-oxadiazole derivatives (**C-4** and **C-5**, **Figure 18**). This means that the corrosive environments can influence the inhibition efficiency of oxadiazole compounds [90]. On the other hand, the substitution of a small carbon chain (e.g., ethyl in the case of **C-2**, **Figure 18**) by another one with the bigger size (e.g., propyl in the case of **C-3**, **Figure 18**) cannot usually induce an enhancement of the inhibition ability of oxazole-based inhibitors. The SEM images of mild steel surface in **Figure 19** confirm the efficacy of **C-2** derivative as an effective corrosion inhibitor.

A novel synthesizing procedure of 3,5-disubstituted 1,2,4-oxadiazole molecule was proposed by Outirite et al. [129]. By means of this procedure, three new 1,2,4-oxadiazole derivatives with pyridinium substituents (**Figure 20**) have been synthesized and reported as excellent corrosion inhibitors for C38 carbon steel in 1 M hydrochloric acid solution [130]. It is well-known that an increase of inhibitor concentration in the corrosive medium mainly leads to an enhancement of its prevention activity. Under this fact, the inhibition capacity of the latest listed derivatives was elevated by raising their amount in considered corrosive solution. On the other hand, the position of nitrogen atoms in pyridine substituents was shown not to have a notable influence on the anti-corrosion property of evaluated compounds. Nevertheless, a remarkable inhibition efficiency of around 95% was obtained at $8 \times 10^{-4}$ M of synthesized 1,2,4-oxadiazole derivatives.

Several isoxazole-based molecules have also demonstrated noticeable protective performance for various metallic materials, such as $Cu_{90}Ni_{10}$ alloy and galvanized and mild steels, under different operating conditions. For instance, two new 5-phenylisoxazole derivatives have been developed and evaluated by Dominguez-Crespo et al. (**C-1** and **C-2**, **Figure 21**) [131]. According to experimental tests, 5-phenylisoxazole compounds exhibited great prevention effectiveness toward the degradation of galvanized steel and copper/nickel alloy. At 5 ppm as inhibitor concentration, the recorded prevention percentages are 100 and 93% for **C-2** and **C-1** in the case of galvanized steel, while in the case of $Cu_{90}Ni_{10}$ alloy they are 88 and 68% for **C-2** and **C-1** compounds. It is interesting to underline that achieved protection efficiencies are comparable to those of the commercial inhibitors (working under the same conditions). Another isoxazole derivative has been found to be an adequate inhibitor for mild steel in 1 M HCl aggressive solution [132]. The

| Oxazole derivative | Metal | Medium | Inhibition efficiency (IE) |
|---|---|---|---|
| 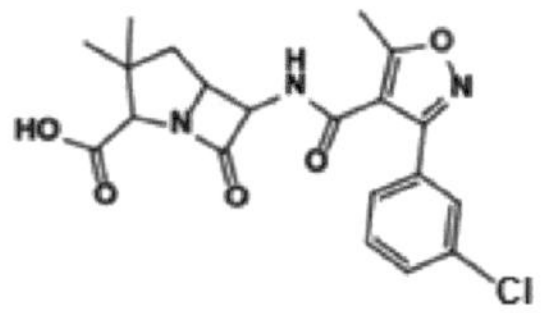 | Mild steel | 0.1 M $H_2SO_4$ | **IE = 94%,** IE was enhanced by synergistic combination with halides at 303 K and $12\times10^{-4}$ M |
| 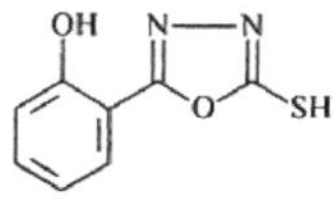 **C-1** 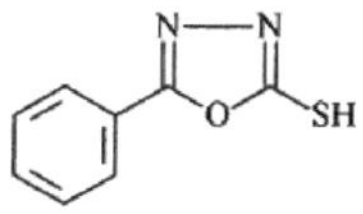 **C-2** 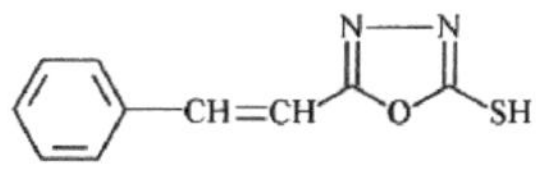 **C-3** | Mild steel | 1.0 M $H_2SO_4$ and 1.0 M HCl | **IE:** C-1 (**78%**) < C-2 (**86%**) < C-3 (**89%**) in 1.0 M $H_2SO_4$ solution, **IE:** C-1 (**71%**) < C-2 (**77%**) < C-3 (**83%**) in 1.0 M HCl solution at 301 ± 2 K and 500 ppm |
| 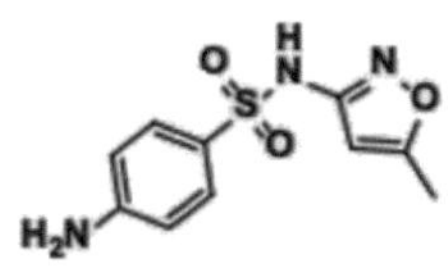 | Mild steel | 1.0 M HCl and 0.5 M $H_2SO_4$ | **IE = 93%** in 1.0 M HCl solution, **IE = 82%** in 0.5 M $H_2SO_4$ solution at 298±1 K and $5\times10^{-3}$ M |

**Table 3.**
*Relevant data related to the application of some oxazole-based compounds as corrosion inhibitors for mild steel.*

molecular structure of the new synthesized derivative (**C-3**) is depicted in **Figure 21**. Both experimental and theoretical approaches pointed out that evaluated **C-3** derivative acts as an effective corrosion inhibitor, in which its inhibition performance reaches 93% at $10^{-3}$ M.

Rather than employing oxazole derivatives, another novel strategy to enhance the anti-corrosion activity of these compounds is the use of its metal complexes. In the recent work, Najeeb [133] has reported the good performance of some metal complexes of a 1,3,4-oxadiazole derivative (**C-6, Figure 18**) against the corrosion of mild steel in 1 M $HNO_3$ medium. As core metal ions, Najeeb has tested Co(II), Ni (II), and Cu(II) ions. As a major outcome of this work, an increase of inhibition efficiency was observed via the metallic complexing process, and the following order of the inhibition efficiency is outlined: Co(II)-oxadiazole > Ni(II)-oxadiazole > Cu(II)-oxadiazole > oxadiazole. Moreover, the inhibition performance of these heterocyclic oxygen/nitrogen compounds can be synergistically enhanced by adding halide ions into the inhibition systems [134].

As was revealed in the literature, many other oxazole-based derivatives have recently stated as good anti-corrosion compounds for several metallic materials that immersed in different corrosive environments. **Table 3** illustrates the relevant data related to the use of some oxazole-based compounds as corrosion inhibitors [134–136].

## 5. Conclusion

In the current chapter, we focused on the application of azole-based compounds as inhibitor agents against metallic corrosion. Almost N-, N&S-, and N&O-azole-containing compounds were found to provide good protection property for numerous metal (or alloy)/medium systems. In this context, three main strategies were adopted to enhance the capability of these compounds to inhibit the corrosion. The first one is based on the synergistic effect, in which supplementary additives (e.g., halide ions) are added into the corrosive media containing azole-based compound, while the chemical modification of azole molecular structures is the second strategy. The latest one is widely used and aimed to introduce further active sites of adsorption within these heterocyclic molecules. Recently, the metallic complex of azole compounds was also reported as an effective strategy to improve their prevention capacity. It is important to outline that N-azole compounds are extensively studied and reported as inhibitors for many metal/medium combinations in comparison with N&S- and N&O-azole ones. Consequently, more attention should be directed to examine the latest two-azole classes, especially oxazole-based compounds.

## Author details

Brahim El Ibrahimi[1] and Lei Guo[2]*

1 Applied Chemistry-Physic Team, Faculty of Sciences, University of IBN ZOHR, Agadir, Morocco

2 School of Material and Chemical Engineering, Tongren University, Tongren, China

*Address all correspondence to: cqglei@163.com

## References

[1] Sastri VS, Ghali E, Elboujdaini M. Corrosion Prevention and Protection Practical Solutions. 1st ed. USA: John Wiley & Sons Ltd.; 2007

[2] Koch GH, Thompson NG, Moghissi O, Payer JH, Varney J. IMPACT (International Measures of Prevention, Application, and Economics of Corrosion Technologies Study). NACE International: Houston, TX; 2016

[3] Kharbach Y, Qachchachi FZ, Haoudi A, Tourabi M, Zarrouk A, Jama C, et al. Anticorrosion performance of three newly synthesized isatin derivatives on carbon steel in hydrochloric acid pickling environment: Electrochemical, surface and theoretical studies. Journal of Molecular Liquids. 2017;**246**:302-316

[4] Ech-chihbi E, Belghiti ME, Salim R, Oudda H, Taleb M, Benchat N, et al. Experimental and computational studies on the inhibition performance of the organic compound “2-phenylimidazo [1,2-a]pyrimidine-3-carbaldehyde” against the corrosion of carbon steel in 1.0 M HCl solution. Surfaces and Interfaces. 2017;**9**:206-217

[5] Ramezanzadeh M, Sanaei Z, Bahlakeh G, Ramezanzadeh B. Highly effective inhibition of mild steel corrosion in 3.5% NaCl solution by green nettle leaves extract and synergistic effect of eco-friendly cerium nitrate additive: Experimental, MD simulation and QM investigations. Journal of Molecular Liquids. 2018;**256**: 67-83

[6] Palanisamy G. Corrosion Inhibitors. In: Singh A, editor. Corrosion Inhibitors. Rijeka: IntechOpen. 2019. DOI: 10.5772/intechopen.80542. Available from: https://www.intechopen.com/books/corrosion-inhibitors/corrosion-inhibitors

[7] El Ibrahimi B, El Mouaden K, Jmiai A, Baddouh A, El Issami S, Bazzi L, et al. Understanding the influence of solution's pH on the corrosion of tin in saline solution containing functional amino acids using electrochemical techniques and molecular modeling. Surfaces and Interfaces. 2019;**17**:100343

[8] Tourabi M, Nohair K, Traisnel M, Jama C, Bentiss F. Electrochemical and XPS studies of the corrosion inhibition of carbon steel in hydrochloric acid pickling solutions by 3,5-bis(2-thienylmethyl)-4-amino-1,2,4-triazole. Corrosion Science. 2013;**75**:123-133

[9] Verma C, Obot IB, Bahadur I, Sherif E-SM, Ebenso EE. Choline based ionic liquids as sustainable corrosion inhibitors on mild steel surface in acidic medium: Gravimetric, electrochemical, surface morphology, DFT and Monte Carlo simulation studies. Applied Surface Science. 2018;**457**:134-149

[10] El Ibrahimi B, Jmiai A, Bazzi L, El Issami S. Amino acids and their derivatives as corrosion inhibitors for metals and alloys. Arabian Journal of Chemistry. 2020;**13**:740-771

[11] El Ibrahimi B, Jmiai A, El Mouaden K, Baddouh A, El Issami S, Bazzi L, et al. Effect of solution's pH and molecular structure of three linear α-amino acids on the corrosion of tin in salt solution: A combined experimental and theoretical approach. Journal of Molecular Structure. 2019;**1196**:105-118

[12] Eicher T, Hauptmann S. The Chemistry of Heterocycles: Structure, Reactions, Synthesis, and Applications. 2nd ed. USA: John Wiley & Sons; 2003

[13] Huang W, Zhao J. Adsorption of quaternary ammonium gemini surfactants on zinc and the inhibitive

effect on zinc corrosion in vitriolic solution. Colloids and Surfaces A: Physicochem. Engineering Aspects. 2006;**278**:246-251

[14] Haleem SMAE, Aal EEAE. Thermometric behaviour of cobalt in $HNO_3$ solutions. Journal of Materials Processing Technology. 2008;**204**: 139-146

[15] Haleem SMAE, Wanees SAE, Aal EEAE, Farouk A. Factors affecting the corrosion behaviour of aluminium in acid solutions. I. Nitrogen and/or Sulphur-containing organic compounds as corrosion inhibitors for Al in HCl solutions. Corrosion Science. 2013;**68**: 1-13

[16] Obot IB, Obi-Egbedi NO, Umoren SA. The synergistic inhibitive effect and some quantum chemical parameters of 2,3-diaminonaphthalene and iodide ions on the hydrochloric acid corrosion of aluminium. Corrosion Science. 2009;**51**:276-282

[17] Fawzy A, Farghaly TA, El-Ghamry HA, Bawazeer TM. Investigation of the inhibition efficiencies of novel synthesized cobalt complexes of 1,3,4-thiadiazolethiosemicarbazone derivatives for the acidic corrosion of carbon steel. Journal of Molecular Structure. 2020;**1203**:127447

[18] Liao LL, Mo S, Luo HQ, Li NB. Corrosion protection for mild steel by extract from the waste of lychee fruit in HCl solution: Experimental and theoretical studies. Journal of Colloid and Interface Science. 2018;**520**: 41-49

[19] Fouda AS, Ismail MA, EL-Ewady GY, Abousalem AS. Evaluation of 4-amidinophenyl-2,2′-bithiophene and its aza-analogue as novel corrosion inhibitors for CS in acidic media: Experimental and theoretical study. Journal of Molecular Liquids. 2017;**240**: 372-388

[20] Sharma J, Ahmad S, Alam MS. Bioactive triazoles: A potential review. Journal of Chemical and Pharmaceutical Research. 2012;**4**:5157-5164

[21] Peyton LR, Gallagher S, Hashemzadeh M. Triazole antifungals: A review. Drugs Today. 2015;**51**:705-718

[22] Sherif ESM, Shamy AME, Ramla MM, Nazhawy AOHE. 5-(Phenyl)-4H-1,2,4-triazole-3-thiol as a corrosion inhibitor for copper in 3.5% NaCl solutions. Materials Chemistry and Physics. 2007;**102**:231-239

[23] Soumoue A, Ibrahimi BE, Issami SE, Bazzi L. Some triazolic compounds as corrosion inhibitors for copper in sulphuric acid. International Journal of Scientific Research. 2014;**3**:349-354

[24] Negrón-Silva GE, González-Olvera R, Angeles-Beltrán D, et al. Synthesis of new 1,2,3-triazole derivatives of uracil and thymine with potential inhibitory activity against acidic corrosion of steels. Molecules. 2013;**18**:4613-4627

[25] Bentiss F, Jama C, Mernari B, Attari HE, Kadi LE, Lebrinib M, et al. Corrosion control of mild steel using 3,5-bis(4-methoxyphenyl)-4-amino-1,2,4-triazole in normal hydrochloric acid medium. Corrosion Science. 2009; **51**:1628-1635

[26] Bentiss F, Bouanis M, Mernari B, Traisnel M, Vezin H, Lagrenée M. Understanding the adsorption of 4H-1,2,4-triazole derivatives on mild steel surface in molar hydrochloric acid. Applied Surface Science. 2007;**253**: 3696-3704

[27] Zheludkevich ML, Yasakau KA, Poznyak SK, Ferreira MGS. Triazole and thiazole derivatives as corrosion inhibitors for AA2024 aluminium alloy. Corrosion Science. 2005;**47**:3368-3383

[28] Issami SE, Bazzi L, Mihit M, Hilali M, Salghi R, Addi EA. Corrosion

inhibition of 70Cu–30Zn alloy in 0.5 M $HNO_3$ by 3-amino-1,2,4-triazole. Journal de Physique IV France. 2005; **123**:307-311

[29] Sherif ESM. Electrochemical investigations on the corrosion inhibition of aluminum by 3-amino-1,2,4-triazole-5-thiol in naturally aerated stagnant seawater. Journal of Industrial and Engineering Chemistry. 2013;**19**:1884-1889

[30] Belghiti ME, Karzazi Y, Dafali A, Obot IB, Ebenso EE, Emran KM, et al. Anti-corrosive properties of 4-amino-3,5-bis(disubstituted)-1,2,4-triazole derivatives on mild steel corrosion in 2 M $H_3PO_4$ solution: Experimental and theoretical studies. Journal of Molecular Liquids. 2016;**216**:874-886

[31] Liu X, Chen S, Zhai H, Shen L, Zhou J, Wu L. The study of self-assembled films of triazole on iron electrodes using electrochemical methods, XPS, SEM and molecular simulation. Electrochemistry Communications. 2007;**9**:813-819

[32] Swathi NP, Alva VDP, Samshuddin S. A review on 1,2,4-triazole derivatives as corrosion inhibitors. Journal of Bio- and Tribo-Corrosion. 2017;**3**:42

[33] Resende GO, Teixeira SF, Figueiredo IF, Godoy AA, Lougon DJF, Cotrim BA, et al. Synthesis of 1,2,3-triazole derivatives and its evaluation as corrosion inhibitors for carbon steel. International Journal of Electrochemistry. 2019;**2019**:6759478

[34] Resende GO, Teixeira SF, Figueiredo IF, et al. Synthesis of 1,2,3-triazole derivatives and its evaluation as corrosion inhibitors for carbon steel. International Journal of Electrochemistry. 2019;**2019**:1-12

[35] Fernandes CM, Alvarez LX, Santos NEd, et al. Green synthesis of 1-benzyl-4-phenyl-1H-1,2,3-triazole, its application as corrosion inhibitor for mild steel in acidic medium and new approach of classical electrochemical analyses. Corrosion Science. 2019;**149**:185-194

[36] Ma Q, Qi S, He X, et al. 1,2,3-Triazole derivatives as corrosion inhibitors for mild steel in acidic medium: Experimental and computational chemistry studies. Corrosion Science. 2017;**129**:91-101

[37] Deng Q, Shi H-W, Ding N-N, et al. Novel triazolylbis-amino acid derivatives readily synthesized via click chemistry as potential corrosion inhibitors for mild steel in HCl. Corrosion Science. 2012;**57**:220-227

[38] Finšgar M, Milošev I. Inhibition of copper corrosion by 1,2,3-benzotriazole: A review. Corrosion Science. 2010;**52**: 2737-2749

[39] Ravichandran R, Rajendran N. Electrochemical behaviour of brass in artificial seawater: Effect of organic inhibitors. Applied Surface Science. 2005;**241**:449-458

[40] Kosec T, Miloseva I, Pihlar B. Benzotriazole as an inhibitor of brass corrosion in chloride solution. Applied Surface Science. 2007;**253**:8863-8873

[41] Zhang D, Gao L, Zhou G. Inhibition of copper corrosion by bis-(1-benzotriazolymethylene)-(2, 5-thiadiazoly)-disulfide in chloride media. Applied Surface Science. 2004;**225**: 287-293

[42] Solehudin A. Performance of benzotriazole as corrosion inhibitors of carbon steel in chloride solution containing hydrogen sulfide. International Refereed Journal of Engineering and Science. 2012;**1**:21-26

[43] Ravichandran R, Nanjundan S, Rajendran N. Corrosion inhibition of brass by benzotriazole derivatives in

NaCl solution. Anti-Corrosion Methods and Materials. 2005;**52**:226-232

[44] Huang H, Wang Z, Gong Y, Gao F, Luo Z, Zhang S, et al. Water soluble corrosion inhibitors for copper in 3.5 wt % sodium chloride solution. Corrosion Science. 2017;**123**:339-350

[45] Sherif EM. Effects of 2-amino-5-(ethylthio)-1,3,4-thiadiazole on copper corrosion as a corrosion inhibitor in 3% NaCl solutions. Applied Surface Science. 2006;**252**:8615-8623

[46] Zhang P, Zhu Q, Su Q, Guo B, Cheng S. Corrosion behavior of T2 copper in 3.5% sodium chloride solution treated by rotating electromagnetic field. Transactions of Nonferrous Metals Society of China. 2016;**26**:1439-1446

[47] Finsgar M, Lesar A, Kokalj A, Milosev I. A comparative electrochemical and quantum chemical calculation study of BTAH and BTAOH as copper corrosion inhibitors in near neutral chloride solution. Electrochimica Acta. 2008;**53**:8287-8297

[48] Umoren SA, Solomon MM. Effect of halide ions on the corrosion inhibition efficiency of different organic species—A review. Journal of Industrial and Engineering Chemistry. 2015;**21**:81-100

[49] Umoren SA, Solomon MM. Synergistic corrosion inhibition effect of metal cations and mixtures of organic compounds: A review. Journal of Environmental Chemical Engineering. 2017;**5**:246-273

[50] Bokati KS, Dehghanian C, Yari S. Corrosion inhibition of copper, mild steel and galvanically coupled copper-mild steel in artificial sea water in presence of 1H-benzotriazole, sodium molybdate and sodium phosphate. Corrosion Science. 2016;**126**:272-285

[51] Kallip S, Bastos AC, Yasakau KA, Zheludkevich ML, Ferreira MGS. Synergistic corrosion inhibition on galvanically coupled metallic materials. Electrochemistry Communications. 2012;**20**:101-104

[52] Coelho LB, Mouanga M, Druart M-E, Recloux I, Cossement D, Olivier M-G. A SVET study of the inhibitive effects of benzotriazole and cerium chloride solely and combined on an aluminium/copper galvanic coupling model. Corrosion Science. 2016;**110**:143-156

[53] Otmacic H, Stupnisek-Lisac E. Copper corrosion inhibitors in near neutral media. Electrochimica Acta. 2003;**48**:985-991

[54] Hmamou DB, Salghi R, Zarrouk A, Aouad MR, Benali O, Zarrok H, et al. Weight loss, electrochemical, quantum chemical calculation, and molecular dynamics simulation studies on 2-(benzylthio)-1,4,5-triphenyl-1H-imidazole as an inhibitor for carbon steel corrosion in hydrochloric acid. Industrial & Engineering Chemistry Research. 2013;**52**:14315

[55] Murulana LC, Singh AK, Shukla SK, Kabanda MM, Ebenso EE. Experimental and quantum chemical studies of some bis (trifluoromethyl-sulfonyl) imide imidazolium-based ionic liquids as corrosion inhibitors for mild steel in hydrochloric acid solution. Industrial and Engineering Chemistry Research. 2012;**51**:13282

[56] Curkovic HO, Stupnisek-Lisac E, Takenouti H. Electrochemical quartz crystal microbalance and electrochemical impedance spectroscopy study of copper corrosion inhibition by imidazoles. Corrosion Science. 2009;**51**:2342-2348

[57] Curkovic HO, Lisac ES, Takenouti H. The influence of pH value on the efficiency of imidazole based corrosion inhibitors of copper. Corrosion Science. 2010;**52**:398-405

[58] Lee W-J. Inhibiting effects of imidazole on copper corrosion in 1 M $HNO_3$ solution. Materials Science and Engineering. 2003;**348**:217

[59] Zhang D-Q, Gao L-X, Zhou G-D. Inhibition of copper corrosion in aerated hydrochloric acid solution by heterocyclic compounds containing a mercapto group. Corrosion Science. 2004;**46**:3031-3040

[60] Zhang D-Q, Gao L-X, Zhou G-D. Synergistic effect of 2-mercapto benzimidazole and KI on copper corrosion inhibition in aerated sulfuric acid solution. Journal of Applied Electrochemistry. 2003;**33**:361

[61] Stupnisek-Lisac E, Gazivoda A, Madzarac M. Evaluation of non-toxic corrosion inhibitors for copper in sulphuric acid. Electrochimica Acta. 2002;**47**:4189-4194

[62] Stupnisek-Lisac E, Bozic AL, Cafuk I. Low-toxicity copper corrosion inhibitors. Corrosion. 1998;**54**:713-720

[63] Gašparac R, Stupnišek-Lisac E. Corrosion protection on copper by imidazole and its derivatives. Corrosion. 1999;**55**:1031-1039

[64] Subramanian R, Lakshminarayanan V. Effect of adsorption of some azoles on copper passivation in alkaline medium. Corrosion Science. 2002;**44**:535-554

[65] Bansal Y, Silakari O. The therapeutic journey of benzimidazoles: A review. European Journal of Medicinal Chemistry. 2012;**20**:6208-6236

[66] Marinescu M. Recent advances in the use of benzimidazoles as corrosion inhibitors. BMC Chemistry. 2019;**13**:136

[67] Obot IB, Edouk UM. Benzimidazole: Small planar molecule with diverse anti-corrosion potentials. Journal of Molecular Liquids. 2017;**246**:66-90

[68] Lewis G. The corrosion inhibition of copper by benzimidazole. Corrosion Science. 1982;**22**:579-584

[69] Moreira RR, Soares TF, Ribeiro J. Electrochemical investigation of corrosion on AISI 316 stainless steel and AISI 1010 carbon steel: Study of the behaviour of imidazole and benzimidazole as corrosion inhibitors. Advances in Chemical Engineering and Science. 2014;**4**:503-514

[70] Guttierez E, Rodriguez JA, Cruz-Borboll J, Alvarado-Rodriguez JG, Thangarasu P. Development of a predictive model for corrosion inhibition of carbon steel by imidazole and benzimidazole derivatives. Corrosion Science. 2016;**108**:23-35

[71] Thang Y, Zhang F, Hu S, Cao Z, Wu Z, Jing W. Novel benzimidazole derivatives as corrosion inhibitors of mild steel in the acidic media. Part I: Gravimetric, electrochemical, SEM and XPS studies. Corrosion Science. 2013;**74**: 271-282

[72] Ramya K, Mohan R, Joseph A. Interaction of benzimidazoles and benzotriazole: Its corrosion protection properties on mild steel in hydrochloric acid. Journal of Materials Engineering and Performance. 2014;**23**:4089-4101

[73] Zhang D, Tang Y, Qi S, Dong D, Cang H, Lu G. The inhibition performance of long-chain alkyl-substituted benzimidazole derivatives for corrosion of mild steel in HCl. Corrosion Science. 2016;**102**:517-522

[74] El-Hajjaji F, Merimi I, Ouasif LE, Ghoul ME, Achour R, Hammouti B, et al. 1-Octyl-2-(octylthio)-1H-benzimidazole as a new and effective corrosion inhibitor for carbon steel in 1 M HCl. Portugaliae Electrochimica Acta. 2019;**37**:131-145

[75] Khaled KF. The inhibition of benzimidazole derivatives on corrosion

of iron in 1 M HCl solutions. Electrochimica Acta. 2003;**48**:2493-2503

[76] Bereket G, Binarbasi A, Ogretir C. Benzimidazole-2-tione and benzoxazole-2-tione derivatives as corrosion inhibitors for aluminium in hydrochloric acid. Anti-Corrosion Methods and Materials. 2004;**51**:282-293

[77] Niamien PM, Kouasi HA, Trokoueri A, Essy FK, Sissouma D, Bokra Y. Copper corrosion inhibition in 1 M $HNO_3$ by two benzimidazole derivatives. International Scholarly Research Notices. **2012**:623754. Available from: https://www.hindawi.com/journals/isrn/2012/623754/

[78] Shanbag AV, Venkatesha TV, Praveen M. Benzimidazole derivatives as corrosion inhibitors for zinc in acid solution. Protection of Metals and Physical Chemistry. 2013;**49**:587-590

[79] Suaad MH, Al-Majidi E, Uday HR, Al-Jeilawi E, Khulood AS, Al-Saadie E. Synthesis and characterization of some 2-sulphanyl benzimidazole derivatives and study of effect as corrosion inhibitors for carbon steel in sulfuric acid solution. Iraqi Journal of Science. 2013;**54**:789-802

[80] Ranjana MM, Nandi MM. Corrosion inhibition of brass in presence of sulphonamidoimidazoline and hydropyrimidine in chloride solution. Indian Journal of Chemical Technology. 2009;**16**:221-227

[81] Popova A, Sokolova E, Raicheva S, Christov M. AC and DC study of the temperature effect on mild steel corrosion in acid media in the presence of benzimidazole derivatives. Corrosion Science. 2003;**45**:33-58

[82] Dutta A, Panja SS, Nandi MM, Sukul D. Effect of optimized structure and electronic properties of some benzimidazole derivatives on corrosion inhibition of mild steel in hydrochloric acid medium: Electrochemical and theoretical studies. Journal of Chemical Sciences. 2015;**127**:921-929

[83] Zhang F, Tang Y, Cao Z, Jing W, Wu Z, Chen Y. Performance and theoretical study on corrosion inhibition of 2-(4-pyridyl)-benzimidazole for mild steel in hydrochloric acid. Corrosion Science. 2012;**61**:1-9

[84] Lgaz H, Salghi R, Jodeh S. Corrosion inhibition potentiality of some benzimidazole derivatives for mild steel in hydrochloric acid: Electrochemical and weight loss studies. International Journal of Corrosion and Scale Inhibition. 2016;**5**:347-359

[85] Rodriguez-Clemente E, Victoria BP, Cervantes-Huevas H, Aldana-Gonzales J, Uruchurtu-Chavarin J, Romero-Romo M, et al. New 1-(2-pyridinyl)-2-(o-, m-, p-hydroxyphenyl) benzimidazoles as corrosion inhibitors for API 5L X52 steel in acid media. Anti-Corrosion Methods and Materials. 2018;**65**:166-175

[86] Ozbay S, Yanardag T, Dincer A, Aksut AA. Benzimidazole Schiff bases as corrosion inhibitors for copper and brass. European International Journal of Science and Technology. 2014;**3**:1-6

[87] Yadav M, Behera D, Kumar S, Sinha RR. Experimental and quantum chemical studies on the corrosion inhibition performance of benzimidazole derivatives for mild steel in HCl. Industrial and Engineering Chemistry Research. 2013;**52**:6318-6328

[88] Yadav M, Kumar S, Purkait T, Olasunkanmi LO, Bahadur I, Ebenso EE. Electrochemical, thermodynamic and quantum chemical studies of synthesized benzimidazole derivatives as corrosion inhibitors for N80 steel in hydrochloric acid. Journal of Molecular Liquids. 2016;**213**:122-138

[89] Wang X, Wan Y, Zeng Y, Gu Y. Investigation of benzimidazole compound as a novel corrosion inhibitor

for mild steel in hydrochloric acid solution. International Journal of Electrochemical Science. 2012;7: 2403-2415

[90] Ahamad I, Quraishi M. Bis (benzimidazol-2-yl) disulphide: An efficient water soluble inhibitor for corrosion of mild steel in acid media. Corrosion Science. 2009;**51**:2006

[91] Dutta A, Saha SK, Banerjee P, Sukul D. Correlating electronic structure with corrosion inhibition potentiality of some bis-benzimidazole derivatives for mild steel in hydrochloric acid: Combined experimental and theoretical studies. Corrosion Science. 2015;**98**:541-550

[92] Kovačević N, Kokalj A. Chemistry of the interaction between azole type corrosion inhibitor molecules and metal surfaces. Materials Chemistry and Physics. 2013;**137**:331-339

[93] Kovačević N, Milošev I, Kokalj A. The roles of mercapto, benzene, and methyl groups in the corrosion inhibition of imidazoles on copper: II. Inhibitor–copper bonding. Corrosion Science. 2015;**98**:457-470

[94] Kokalj A. Is the analysis of molecular electronic structure of corrosion inhibitors sufficient to predict the trend of their inhibition performance. Electrochimica Acta. 2010;**56**:745-755

[95] Kovačević N, Kokalj A. Analysis of molecular electronic structure of imidazole- and benzimidazole-based inhibitors: A simple recipe for qualitative estimation of chemical hardness. Corrosion Science. 2011;**53**:909-921

[96] Gustincic D, Kokalj A. A DFT study of adsorption of imidazole, triazole, and tetrazole on oxidized copper surfaces: $Cu_2O(111)$ and $Cu_2O(111)$-w/o-$Cu^{CUS}$. Physical Chemistry Chemical Physics. 2015;**17**:28602-28615

[97] El Ibrahimi B, Soumoue A, Jmiai A, Bourzi H, Oukhrib R, El Mouaden K, et al. Computational study of some triazole derivatives (un- and protonated forms) and their copper complexes in corrosion inhibition process. Journal of Molecular Structure. 2016;**1125**:93-102

[98] Quaraishi MA, Rawat J, Ajmal M. Dithiobiurets: A novel class of acid corrosion inhibitors for mild steel. Journal of Applied Electrochemistry. 2000;**30**:745

[99] Wang L. Evaluation of 2-mercaptobenzimidazole as corrosion inhibitor for mild steel in phosphoric acid. Corrosion Science. 2001;**43**:2281

[100] Guo L, Obot IB, Zheng X, Shen X, Qiang Y, Kaya S, et al. Theoretical insight into an empirical rule about organic corrosion inhibitors containing nitrogen, oxygen, and sulfur atoms. Applied Surface Science. 2017;**406**: 301-306

[101] Raviprabha K, Bhat RS. Inhibition effects of ethyl-2-amino-4-methyl-1,3-thiazole-5-carboxylate on the corrosion of AA6061 alloy in hydrochloric acid media. Journal of Failure Analysis and Prevention. 2019;**19**:1464-1474

[102] Tan B, Zhang S, Liu H, Guo Y, Qiang Y, Li W, et al. Corrosion inhibition of X65 steel in sulfuric acid by two food flavorants 2-isobutylthiazole and 1-(1,3-thiazol-2-yl) ethanone as the green environmental corrosion inhibitors: Combination of experimental and theoretical researches. Journal of Colloid and Interface Science. 2019;**538**:519-529

[103] Chugh B, Singh AK, Thakur S, Pani B, Pandey AK, Lgaz H, et al. An exploration about the interaction of mild steel with hydrochloric acid in the presence of N-(benzo[d]thiazole-2-yl)-1-phenylethan-1-imines. Journal of Physical Chemistry C. 2019;**123**: 22897-22917

[104] Xiong L, He Z, Han S, Tang J, Wu Y, Zeng X. Tribological properties study of N-containing heterocyclic imidazoline derivatives as lubricant additives in water glycol. Tribology International. 2016;**104**:98-108

[105] Shukla SK, Quraishi M. Ceftriaxone: A novel corrosion inhibitor for mild steel in hydrochloric acid. Journal of Applied Electrochemistry. 2009;**39**:1517

[106] Mistry BM, Jauhari S. Corrosion inhibition of mild steel in 1 N HCl solution by mercapto-quinoline schiff base. Chemical Engineering Communications. 2014;**201**:961-981

[107] Fouda A, Ellithy A. Inhibition effect of 4-phenylthiazole derivatives on corrosion of 304L stainless steel in HCl solution. Corrosion Science. 2009; **51**:868

[108] Kumar JA, Simant S, Ankur V, Veerasamy R, Kishore AR. 1,3,4-Thiadiazole and its derivatives: A review on recent progress in biological activities. Chemical Biology & Drug Design. 2013;**81**:557-576

[109] Yang H, Cui-Yun L, Xiao-Ming W, Yong-Hua Y, Hai-Liang Z. 1,3,4-Thiadiazole: Synthesis, reactions, and applications in medicinal, agricultural, and materials chemistry. Chemical Reviews. 2014;**114**:5572-5610

[110] Xiong S, Liang D, Ba Z, Zhang Z, Luo S. Adsorption behavior of thiadiazole derivatives as anticorrosion additives on copper oxide surface: Computational and experimental studies. Applied Surface Science. 2019; **492**:399-406

[111] Wei DP, Cao L, Wang LL. An investigation into the antiwear, antioxidation, and anticorrosion behaviour of some derivatives of 2,5-dimercapto-1,3,4-thiadiaxole. Lubrication Science. 1995;7:365-377

[112] Gemmill RM. Corrosion inhibited lubricant composition. US; 1980

[113] Verma C, Quraishi MA, Ebenso EE. Microwave and ultrasound irradiations for the synthesis of environmentally sustainable corrosion inhibitors: An overview. Sustainable Chemistry and Pharmacy. 2018;**10**:134-147

[114] Palomar-Pardavé M, Romero-Romo M, Herrera-Hernández H, Abreu-Quijano M, Likhanova NV, Uruchurtu J, et al. Influence of the alkyl chain length of 2 amino 5 alkyl 1,3,4 thiadiazole compounds on the corrosion inhibition of steel immersed in sulfuric acid solutions. Corrosion Science. 2012;**54**: 231-243

[115] Döner A, Solmaz R, Özcan M, Kardaş G. Experimental and theoretical studies of thiazoles as corrosion inhibitors for mild steel in sulphuric acid solution. Corrosion Science. 2011;**53**: 2902

[116] Salman TA, Zinad DS, Jaber SH, Al-Ghezi M, Mahal A, Takriff MS, et al. Effect of 1,3,4-thiadiazole scaffold on the corrosion inhibition of mild steel in acidic medium: An experimental and computational study. Journal of Bio- and Tribo-Corrosion. 2019;**5**:48

[117] Bawazeer TM, El-Ghamry HA, Farghaly TA, Fawzy A. Novel 1,3,4-thiadiazolethiosemicarbazones derivatives and their divalent cobalt-complexes: Synthesis, characterization and their efficiencies for acidic corrosion inhibition of carbon steel. Journal of Inorganic and Organometallic Polymers and Materials. 2020;**30**: 1609-1620

[118] Zucchi F, Trabanelli G, Monticelli C. The inhibition of copper corrosion in 0.1 M NaCl under heat exchange conditions. Corrosion Science. 1996;**38**:147-154

[119] Szőke AF, Szabó GS, Hórvölgyi Z, Albert E, Végh AG, Zimányi L, et al. Accumulation of 2-acetylamino-5-mercapto-1,3,4-thiadiazole in chitosan coatings for improved anticorrosive effect on zinc. International Journal of Biological Macromolecules. 2020;**142**: 423-431

[120] Alagawadi KR, Mahajanshetti CS, Jalalpure SS. Synthesis of 5-aryl-2-acylthio-1,3,4-oxadiazoles and their antibacterial activity. Indian Journal of Heterocyclic Chemistry. 2005;**14**:315

[121] Tyrkov AG, Sukhenko LT. Synthesis and antimicrobial activity of substituted nitro-1,2,4-oxadiazole-5-carbaldehyde hydrazones. Pharmaceutical Chemistry Journal. 2004;**38**:376

[122] Fouda AS, Elmorsi MA, Fayed T, Said IAE. Oxazole derivatives as corrosion inhibitors for 316L stainless steel in sulfamic acid solutions. Desalination and Water Treatment. 2016;**57**:4371-4385

[123] Rahmani H, El-Hajjaji F, Hallaoui AE, Taleb M, Rais Z, Azzouzi ME, et al. Experimental, quantum chemical studies of oxazole derivatives as corrosion inhibitors on mild steel in molar hydrochloric acid medium. International Journal of Corrosion and Scale Inhibition. 2018;**7**: 509-527

[124] Rugmini Ammal P, Prasad AR, Ramya K, John S, Joseph A. Protection of mild steel in hydrochloric acid using methyl benzimidazole substituted 1,3,4-oxadiazole: Computational, electroanalytical, thermodynamic and kinetic studies. Journal of Adhesion Science and Technology. 2019;**33**: 2227-2249

[125] Ammal PR, Prajila M, Joseph A. Effect of substitution and temperature on the corrosion inhibition properties of benzimidazole bearing 1,3,4-oxadiazoles for mild steel in sulphuric acid: Physicochemical and theoretical studies. Journal of Environmental Chemical Engineering. 2018;**6**:1072-1085

[126] Ammal PR, Prajila M, Joseph A. Effective inhibition of mild steel corrosion in hydrochloric acid using EBIMOT, a 1,3,4-oxadiazole derivative bearing a 2-ethylbenzimidazole moiety: Electro analytical, computational and kinetic studies. Egyptian Journal of Petroleum. 2018;**27**:823-833

[127] Fox PG, Bradely PA. 1:2:4-Triazole as a corrosion inhibitor for copper. Corrosion Science. 1980;**20**:643

[128] Bentiss F, Traisnel M, Lagrenee M. The substituted 1,3,4-oxadiazoles: A new class of corrosion inhibitors of mild steel in acidic media. Corrosion Science. 2000;**42**:127-146

[129] Outirite M, Lebrini M, Lagrenée M, Bentiss F. New one step synthesis of 3,5-disubstituted 1,2,4-oxadiazoles. Journal of Heterocyclic Chemistry. 2007;**44**: 1529

[130] Outirite M, Lagrenée M, Lebrini M, Traisnel M, Jama C, Vezin H, et al. AC impedance, X-ray photoelectron spectroscopy and density functional theory studies of 3,5-bis(n-pyridyl)-1,2,4-oxadiazoles as efficient corrosion inhibitors for carbon steel surface in hydrochloric acid solution. Electrochimica Acta. 2010;**55**:1670-1681

[131] Dominguez-Crespo MA, Zepeda-Vallejo LG, Torres-Huerta AM, Brachetti-Sibaja SB, Palma-Ramirez D, Rodriguez-Salazar AE, et al. New triazole and isoxazole compounds as corrosion inhibitors for Cu-Ni (90/10) alloy and galvanized steel substrates. Metallurgical and Materials Transactions A: Physical Metallurgy and Materials Science. 2020;**51**:1822-1845

[132] Lahmidi S, Elmsellem H, Elyoussfi A, Sebbar NK, Essassi EM,

Ouzidan Y, et al. Investigation of corrosion inhibition of mild steel in 1 M HCl by 3-methyl-4-(3-methyl-isoxazol-5-yl)isoxazol-5(2H)-one monohydrate using experimental and theoretical approaches. Der Pharma Chemica. 2016; **8**:294-303

[133] Najeeb DA. Inhibition efficiency and corrosion rate studies of mild steel in nitric acid using 2-thioacetic acid-5-pyridyl-1,3,4-oxadiazole complexes. International Journal of Corrosion and Scale Inhibition. 2019;**8**:717-725

[134] Eddy NO, Ebenso EE. Adsorption and quantum chemical studies on cloxacillin and halides for the corrosion of mild steel in acidic medium. International Journal of Electrochemical Science. 2010;**5**:731-750

[135] Quraishi MA, Sardar R. Corrosion inhibition of mild steel in acid solutions by some aromatic oxadiazoles. Materials Chemistry and Physics. 2002;**78**:425-431

[136] El-Naggar MM. Corrosion inhibition of mild steel in acidic medium by some sulfa drugs compounds. Corrosion Science. 2007;**49**:2226-2236

Chapter 6

# Azoles for Renewable Energy Development and Wood Treatment

*Nana Derkyi*

## Abstract

Azole applications in energy are empirical and, despite increasing interest in azole and energy research, many challenges remain in synthesizing and processing azoles with functionality for energy applications. The use of azole in wood treatment has been effective to some extent in producing durable wood; however, there is still the need for improving the treatment of wood species. This chapter seeks solutions which are developed systematically with scientific validation principles. Consequently, this chapter aims to provide a concise overview of integrating azoles in materials used for renewable energy processing and applications, and wood treatment, with an outlook on challenges and opportunities.

**Keywords:** azole applications, organic solar cells, polymer electrolyte membranes, fungal decay, wood treatment, functionality, copper azole, overview

## 1. Introduction

There exists a huge demand for favorable new materials in the research space. With this huge demand, the building of a wide array of custom-made materials for different and also multiple applications have been made possible. Most often, the design of conjugated molecules is commonly built around small functionalized aromatic polycyclic systems like azole (**Figure 1**). These systems are built up as pieces of larger conjugated systems like polymers.

In an energy setting that progressively requires efficient and cleaner energy sources, fuel cells are considered as promising electrochemical devices for meeting such demand. This is because they can deliver electric energy with high efficiency and low environmental effect, converting the energy kept in fuels with no pollution. The proton-exchange membrane fuel cells (PEMFC) are known to be one of the most promising sources within the numerous kinds of current fuel cells owing to their great power density and high power-to-weight ratio. One of the downsides of current cells is linked to the electrolytes presently in use, which limit their use to temperatures below 100°C when working with water-assisted proton conduction [1–4]. Operating temperatures above 100°C increase the performance of PEMFC due to a quicker electrode reaction which takes place without carbon monoxide poisoning of the platinum electro-catalyst, high energy efficiency and easier heating, [1–4]. Organic semiconductors (OSCs) have attracted much attention over the past few decades owing to their unique properties, which allow them to be included in a host of electronic device applications.

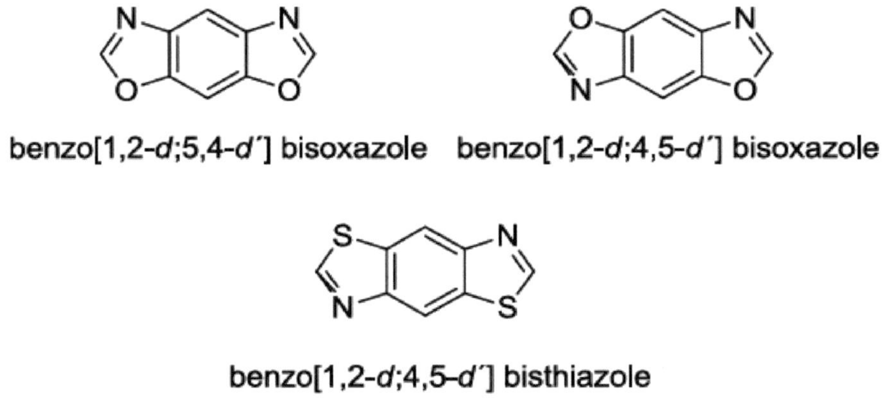


**Figure 1.**
*Basic benzobisazole units.*

Wood has been used as a traditional material for the construction of marine structures, such as groynes and jetties [5], and yachts and other boats. The application of wood in marine structures is attributable to the wood explicit properties. The factors for utilizing wood in marine structures include ease of construction and repair, relatively low energy costs of production, high strength-to-weight ratio, and renewability. However, biodegradation of wood is predominantly harsh in maritime construction due to the action of marine wood borers and crustaceans, in contrast to beetles, decay fungi, and termites active above the waterline [6]. In borer attack prevention, biocides are, in some cases, used to treat wood [6].

The use of durable hardwood species, mostly tropical hardwoods which are resistant to biodegradation, has led to tropical deforestation that continues to be a cause of concern. There are many other factors associated with this tropical deforestation other than the use of timber in maritime structures. The decrease in naturally durable species has necessitated the treatment of softwoods by using preservatives to achieve suitable protection for the wood under service conditions, mainly for outdoor applications [7].

This chapter seeks solutions which are developed systematically with scientific validation principles. Consequently, this chapter aims to provide a concise overview of integrating azoles in materials used for renewable energy processing and wood treatment, with an outlook on challenges and opportunities.

## 2. Chemistry of azoles

Azole compounds are part of a large class of heterocyclic compounds in Organic chemistry. Azoles are five-membered heterocyclic compounds containing a nitrogen atom and at least one other non-carbon atom of either nitrogen, sulfur, or oxygen [8]. They include the heterocyclic rings in **Figure 2**.

The chemistry of 1,2,3-triazoles gained much attention since the discovery of the copper-catalyzed alkyne-azide cycloaddition (CuAAC) reaction, which delivers the 1,4-regioisomer exclusively in high yields [8].

Triazoles are five-membered aromatic heterocyclics, containing three nitrogen atoms. These atoms may be found arranged consecutively or not, given the isomers 1,2,3-triazoles or 1,2,4-triazoles, respectively (**Figure 3**) [8].

The structure of 1,2,3-triazoles, as shown in **Figure 3**, may exist in two diverse tautomers, per their position of the N-H bond on the ring. The position of the N-H

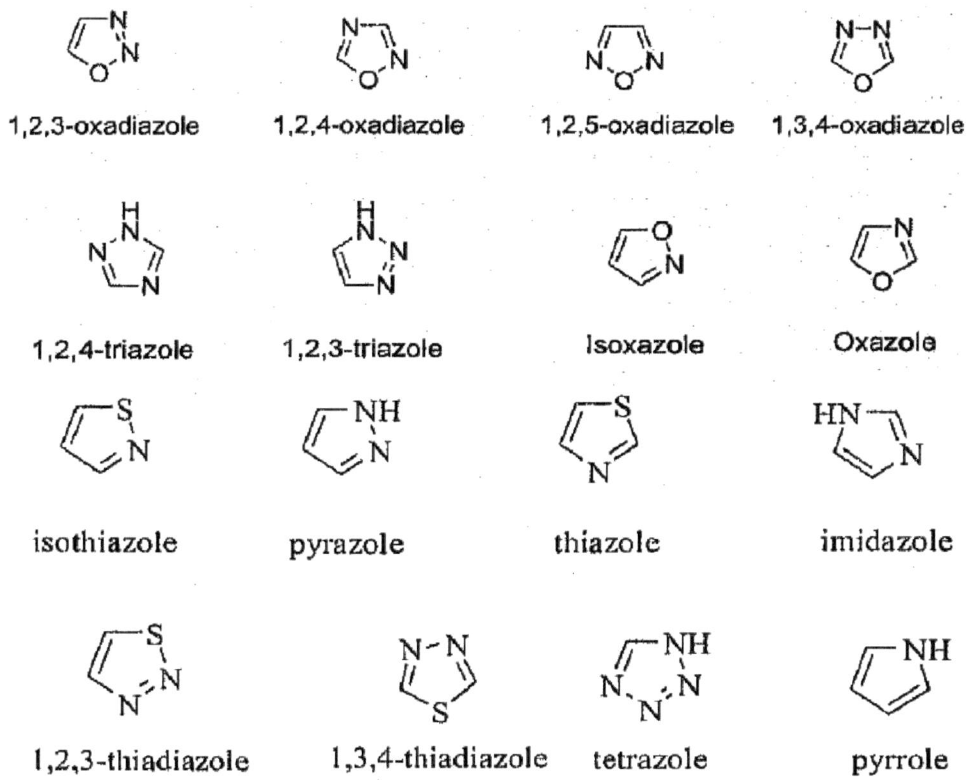


**Figure 2.**
*Examples of azole compounds.*

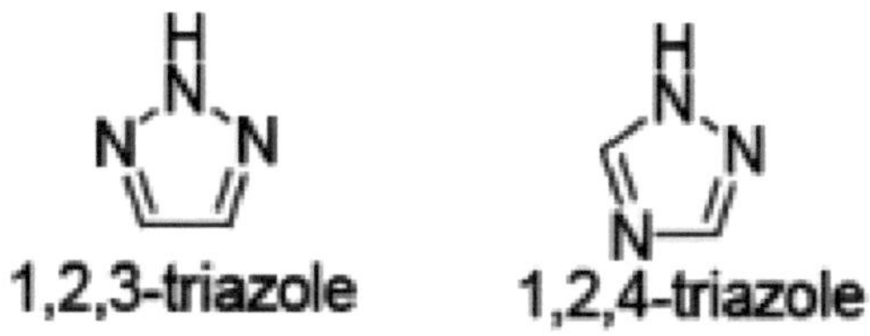


**Figure 3.**
*Structures of isomeric triazoles.*

bond can be on nitrogen 1 (1H) or in nitrogen 2 (2H). If the 1,2,3-triazole skeleton is substituted at the nitrogen and carbon atoms, dissimilar regioisomers may be obtained as shown in **Figure 4**.

When compared to their 1,2,4 isomers, the 1,2,3-triazoles have very distinct properties and more importantly, the N2 substituted has different properties than the N1 as well as N3, despite the structural likeness. For instance, the differences in basicity between the N1 and N2 isomers could be responsible for their different behavior within biological systems [8]. The N2-1,2,3-triazole core is found in several bioactive compounds, including antifungals [8]. The different tautomers shown in **Figure 5** have distinct physical, chemical and biological properties [8].

The substitution of the triazole implies the study of the N3 -H tautomer, in addition to the N1 -H and N2 -H. **Table 1** gives a summary of some energy values obtained by substituting the 1,2,3-triazole at carbon with different substituents, generating tautomers. It has been observed that in all cases studied, the N2 -H tautomer is the most stable. **Figure 6** depicts substituent X on the different

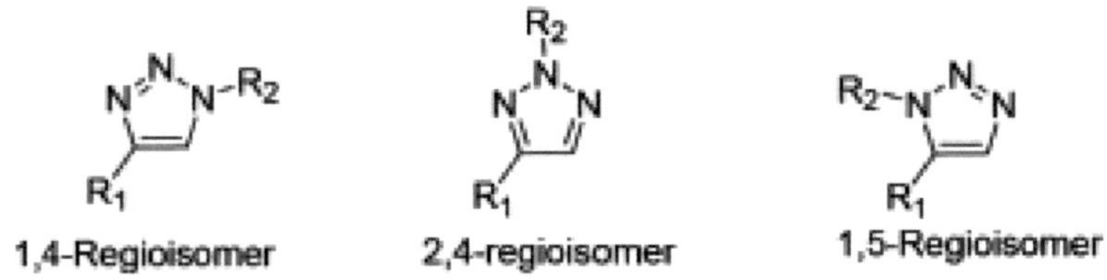


**Figure 4.**
*Substituted 1,2,3-triazoles with dissimilar regioisomers of carbon and nitrogen.*

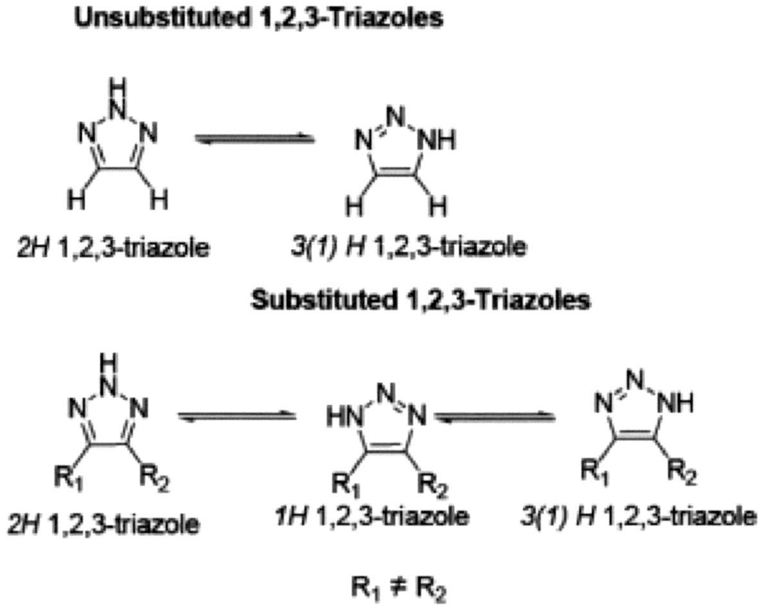


**Figure 5.**
*Different tautomers of substituted and unsubstituted 1,2,3-triazoles [8].*

| X | ΔE (kJ/mol) | | |
|---|---|---|---|
| | $N^3$ | $N^2$ | $N^1$ |
| $NH_2$ | 30.734 | 0 | 22.019 |
| F | 37.059 | 0 | 21.340 |
| Cl | 27,557 | 0 | 20,780 |
| $CH_3$ | 20,566 | 0 | 20,080 |
| CN | 22,475 | 0 | 18,004 |
| $NO_2$ | 22,194 | 0 | 16,941 |
| CHO | 17,188 | 0 | 15,530 |
| COOH | 12,081 | 0 | 13,261 |
| OH | 35.946 | 0 | 18,578 |
| $CONH_2$ | 23,266 | 0 | 10,892 |
| $BF_2$ | 12,336 | 0 | 16,966 |
| $BH_2$ | 5547 | 0 | 15,815 |

**Table 1.**
*The relative energy of N1-H and N3-H to N2-H in the presence of different substituents [9].*

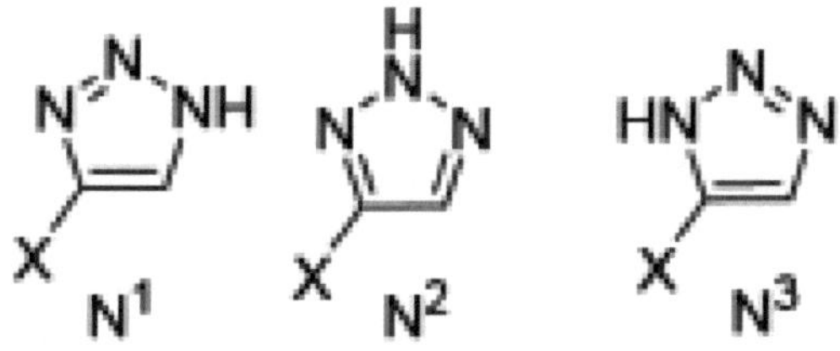

**Figure 6.**
*Substituents on different azole tautomers.*

tautomers, and from **Table 1** it can be realized that the substituent X has a strong correlation on the stabilization of the N1-H tautomer against the N3-H tautomer. It can be observed from **Table 1** that the donor or electron-withdrawing performance of different groups does not render differences in the stabilization of the tautomers. However, the steric hindrance and hydrogen bonds between the substituent and the adjacent azole nitrogen atom which seem to be the preeminent factor in the stabilization of N1-H versus N3-H.

Azoles are known for their broad-spectrum biological activities including antimicrobial, anti-inflammatory, analgesic, antimitotic, anticonvulsive, diuretic and many other uses as main ingredients in many drugs, [10]. Azoles are also known for their usage wood preservatives [11].

Generally, isolated natural products of heterocyclic nature act as lead compounds for the development of new molecules of bioactive interest. Also, most of the heterocyclic compounds are synthesized from readily available fine chemicals. In this respect, synthesis and characterization of new molecular entities incorporating heterocyclic structures are of great importance.

Azoles constitute a crucial category of antifungal agents in clinical, agricultural, and wood treatment uses. In general, they target the inhibition of ergosterol synthesis. Fungi are eukaryotes just like mammalian cells, and so agents that affect protein or nucleic acid biosynthesis are likely to display general eukaryotic toxicity [12]. Ergosterol, the predominant component of fungal cell membranes, is, therefore, evident and specific target for fungal inhibition.

## 3. History of azoles

In the beginning of the development of organic chemistry, heterocyclic chemistry of which azoles belong has held center stage in the development of molecules to enhance the quality of human life. Examples include drugs development, agricultural produce and wood preservation, as well as energy applications [8, 13]. Some of these organic compounds, as early as the 1950s, had been reported to have electrical conductivity [13]. However, the first breakthroughs did not occur until the 1970s. OSCs have been incorporated into various electronic devices, including organic photovoltaics (OPVs) [14, 15], some of which have been developed commercially [16, 17].

Researchers had become interested in the antifungal activity of azole compounds since 1958 after the introduction of topical chlormidazole. However, the first report of the antifungal activity of an azole compound, benzimidazole, was already described in 1944 by Woolley [18]. The initial reports of antifungal properties for imidazoles were published in the late 1960s [19]. During that period, three new topical compounds: clotrimazole, developed by Bayer Ag (Germany), and miconazole and econazole, both developed by Janssen Pharmaceutica (Belgium) were introduced [20].

There have been significant changes in the wood preservation industry over the past several years. For ground contact applications, copper-based systems have replaced the chromate copper arsenate (CCA) product used for many years but had some corrosion and mold issues during the initial phases of the transition [20]. Creosote was virtually the only preservative until various new preservatives were introduced in the 1930 and 1940s. There was a continued evolution of preservative systems, and in 1990, copper-azole preservatives were introduced [20].

## 4. Azole-based materials for renewable energy applications

There is quite a large number of azole compounds that are suitable for energy applications. These include but not limited to imidazoles, diazoles, triazoles, tetrazoles, thyazoles and pyrazoles [2]. A fuel cell, which is a device that provides electrical energy with high efficiency and low environmental impact, converts the chemical energy stored in fuel, such as hydrogen, methanol, ethanol, etc., directly and efficiently to electrical energy. In fuel cell applications, the fuel cell consists of thermal, humidification, and reactant/product management systems, electronics, and the membrane electrode assembly (MEA) [2]. The membrane electrode assembly is the electrochemically active portion of the cell, which contains an ion-permeable but electrically non-conductive electrolyte sandwiched between an anode and cathode at which oxidation of fuel and reduction of oxygen occur. Organic ionic liquids, molten salts, and strong acids/bases can all be utilized as the electrolyte separating anode and cathode. However, to minimize corrosion and gas diffusion and to improve the mechanical strength of fuel cells, solid oxide and polymer electrolyte membranes are the predominant separators in modern systems. Engineering of the complete fuel cell, including its support systems, is, therefore, an essential task in which the materials and chemistry are focused on the electroactive MEA. Thus, the proton-exchange membrane (PEM) is a significant component for the operation of proton-exchange membrane fuel cell (PEMFC).

Generally, PEM are made of polymeric organic compounds containing acidic functions (example, Nafion). The restrictions of modern membranes have nurtured the research and development of alternative membranes, including doped polybenzimidazole (a combination of Nafion and metal oxides), organosiloxane based on inorganic-organic hybrids with various acidic species, and sulfonated polymers based on aromatic hydrocarbons [8–14].

Typically, some amount of hydration is essential to conduct ions, and there are some new materials merging acceptor and donor ion carrier abilities of numerous groups [10]. Heterocyclics do act as a proton-conducting species, due to the amphoteric behavior of nitrogen. Thus, they can be used either as dopant or pendant groups in PEMFC devoid of the need to use external humidification. The properties of numerous heterocyclics, including benzimidazole and triazole, permit them to be used in materials operational above 100°C [21–24].

Several azole derivatives have been synthesized to become precursors for novel fuel cell membrane materials. For example, from the azole compound 4,7-dibromobenzimidazole, new phosphonate-, hydroxybisphosphonate- and aminobisphosphonate benzimidazole derivatives substituted at N-1 position have been synthesized in good yields. Again, new regioisomers of phosphonate- and aminobisphosphonatebenzotriazole derivatives substituted at N-1 or N-2 positions have been synthesized in good yields from 4,7-dibromobenzotriazole. Characterization by NMR, IR spectroscopy and mass spectrometry (low and high resolution) of these compounds have been fully done allowing the assignment of regioisomers [24].

Considering chemical viewpoint, two principal structural features give rise to the properties of organic semiconductors (OSCs). These are conjugated core or support, and countless types of solubilizing side chains. These two structural features, exact by chemical synthesis, have a wide influence on the nano-scale morphology, optoelectronic energy levels, and the bulk physical features of these materials.

Standard monomer building blocks to construct conjugated polymer for solar cells include azoles, and they are categorized by the number of rings and way of linking. Polybenzobisazoles, a class of polymers that are known for their exceptional thermal stability and high tensile strength of fibers spun from them, are important in organic solar cells applications. 9 An example is poly(*p*-phenylene-2,6-benzobisoxazole) which is a liquid crystalline polymer based on benzo[1,2-*d*; 4,5-*d'*]benzobisoxazole that is spun into fibers commercially sold under the name Zylon® [25]. Because of the previous use of polybenzobisazoles in high-performance applications, all of the necessary monomers can be synthesized on an industrial scale, and purified without the use of column chromatography. This ability to synthesize monomers on an industrial scale is advantageous for large-scale synthesis. Furthermore, the benzobisazole ring system is electron-deficient and planar, which leads to strong intermolecular interactions and good charge transport properties within polymer films [26].

## 5. Current trends and prospects of azoles in renewable energy development

### 5.1 Fuel cells

The quest for clean and efficient energy has motivated the search for new materials to develop environmentally friendly energy applications. Fuel cells have potential for alternative clean and efficient energy conversion devices with zero pollution [1, 27]. Considering the various kinds of fuel cells, the proton-exchange membrane fuel cells (PEMFCs) are known to be sources of power, due to their inherent great high power-to-weight ratio and power density. The PEM is a key material for the operation of PEMFC. In recent years, PEM has been a focus of many research works, to obtain membranes with good chemical/thermal stability, high proton conductivity, low electrical conductivity, low permeability to fuel and oxidant, good mechanical properties and cost-effective [1, 27]. Temperature is a critical factor in operating PEMFCs due to its correlation with the water content of fuel cell. The processes above 100°C increase the performance of the fuel cell due to quicker electrode reaction without carbon monoxide poisoning of the platinum electrocatalyst, easier heating, water consumption management and high energy efficiency [1, 27]. The PEM is usually made of polymeric organic compounds containing acidic ends; however, the proton transport properties of these membranes are strongly related to the water content and, therefore, limit their operating temperatures up to 90°C [1, 27].

The limitations of temperature have brought about an increased interest in research and development of new alternative membranes. Among them, a variety of membranes have been developed as alternative to the perfluorosulphonic polymers,, such as polybenzimidazole (PBI)-doped composites of Nafion and metal oxides, sulfonated polymers based on aromatic hydrocarbons, and organosiloxane polymers based on inorganic-organic hybrids with various acidic species [1, 27].

Azoles as amphoteric species conduct protons as pure compounds. Recent computational studies are geared towards elucidating the mechanism underlying structural diffusion in some of these amphoteric species. In protonated imidazole, charge transport occurs via a method directly analogous to aqueous transport, exhibiting rapid molecular reorientations and a shifting hydrogen bond network [28].

In several azole-based systems, proton conductivity is being investigated by pulsed field gradient spin-echo nuclear magnetic resonance (PFGSE-NMR) and impedance spectroscopies, to confirm results obtained by computational simulations [29, 30]. Improving the proton conductivity of polymer electrolytes under high temperature and low humidity conditions by imbibing polymers (**Figure 7**) that exhibit favorable hydrated properties with small-molecule amphoteric species is a current research agenda. These molecules effectively replace the water that would traditionally exist as a proton solvent, enabling proton conduction under these dehydrating conditions. Doping studies of PEM with azoles are thus high on the research agenda [31, 32].

## 5.2 Solar cells

Enormous potential exists in solar energy to take the place of fossil fuels due to its vast energy stock and availability worldwide [33]. Solar energy conversion system is traditionally based on silicon technology. However, the wide use of silicon-based solar cell technology is limited by its high power conversion cost [34]. To address this issue, organic solar cell has been developed to replace Si-solar cell [35].

The optoelectronic properties of polymeric semiconductor materials can be used for the fabrication of photonic devices. If key structural requirements are met, these materials exhibit distinctive properties such as solution processability, large charge-transporting capabilities, and/or broad optical absorption. Developments in the area of π-conjugated polymeric semiconductors for bulk-heterojunction photovoltaic cell (BHJ-OPV) or organic solar cell (OSC) applications have been made, and these conjugated polymers (CPs) have become pervasive in photovoltaic cells applications [36].

Conjugated polymers offer several advantages over their inorganic counterparts, including solution processability to reduce fabrication costs, and the ability to tune their properties via organic synthesis, which enables optimization for use in specific applications. Currently, a good and effective strategy for adjusting the optical and electronic properties of conjugated polymers is through the integration of alternating electron-donating and electron-accepting comonomers within the polymer backbone. This approach which is a current trend, has afforded many materials with narrow bandgaps suitable for effective harvesting of solar energy. For example, a synthesis of benzo[1,2-c:4,5-c]bis[1,2,5]thiadiazole containing donor-acceptor monomers and their acid-catalyzed polymerization has been reported [37].

Conjugated polymer-based solar cell (PSC) has several important advantages compared with conventional Si-based solar cell. These include solution processability by spin-coating, ink-jet printing and roll-to-roll processing to reduce manufacturing cost; tunable physical properties, and mechanical flexibility for PSC application on curved surfaces [38].

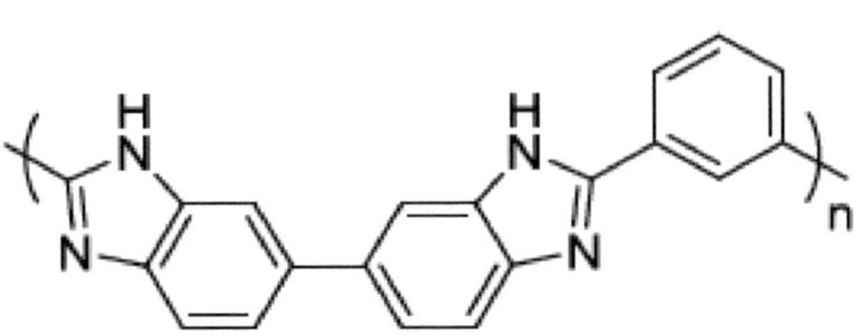


**Figure 7.**
*Chemical structure of unsubstituted polybenzimidazole (PBI) (poly-2,2-m-phenylene-5,5-bibenzimidazole).*

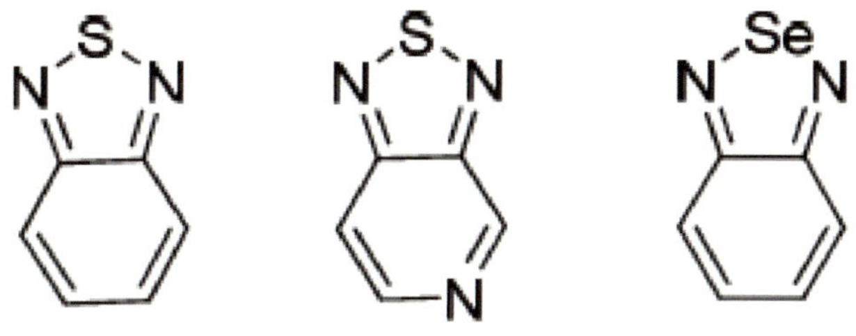


**Figure 8.**
*Structure of (a) BT, (b) aza-BT and (c) Se-BT.*

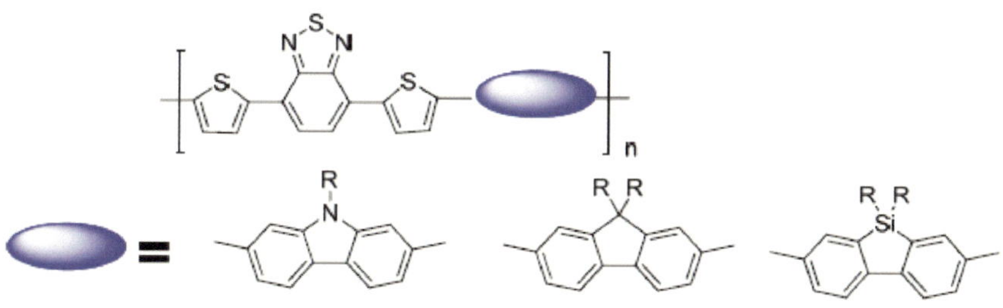

**Figure 9.**
*Benzothiadiazole containing low bandgap polymers.*

An important electron-accepting unit in organic electronics including organic solar cells is 2,1,3-benzothiadiazole (BT). Two types of BT-fused units have been synthesized. These are thiazole-fused BT containing electron-withdrawing C=N bond and imidazole-fused BT containing electron-donating nitrogen atom as well as an electron-withdrawing C=N bond. Theoretical calculations and measurements by electrochemical means suggest that thiazole-fusion enhances electron-accepting ability [39]. In contrast, imidazole-fusion bestows the benzothiadiazole skeleton with electron-donating capacity while sustaining its electron-accepting ability. Besides, in thiazole-fused BT units, the electronic configuration could be additionally controlled by tuning the oxidation number of the sulfur atom in methylthio group at the fused thiazole ring [39]. Thus, the electron withdrawing ability of BT can be further increased by replacing one carbon atom with $sp^2$-hybridized N atom (**Figure 8**).

A variety of low bandgap polymers containing BT have been synthesized and tested for PSC performance (**Figure 9**) [40, 41].

## 6. Challenges of azoles in renewable energy development

Doping of azoles with strong acids can bring about the problems, including incorporating some of the undesirable anhydrous properties of these acids (e.g. high water uptake, physical stiffening of the polymer matrix) into mixed materials. In fuel cells applications, introducing polymer electrolytes with small-molecule proton solvent does not entirely solve the problem of anhydrous proton conductivity. For instance, imidazole still evaporates or is washed from the membrane over time. Also, unlike water, its equilibrium partial pressure in the surrounding atmosphere is negligible, and once it leaves the membrane, it is not easy to re-introduce. Even for species large enough to have little vapor pressure like imidazoles with long tethers (**Figure 10**), leaking along with the water that is formed and expelled during fuel

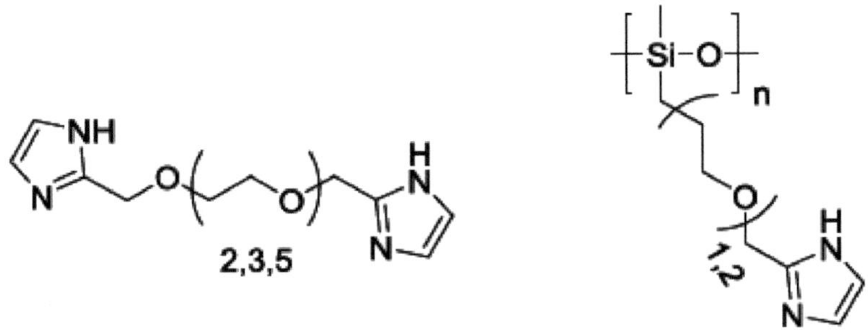


**Figure 10.**
*Tethered imidazoles used to investigate relative rates of a vehicle and structural diffusion in imidazole-based molecules [42].*

cell operation is a significant challenge. This leaking leads to both loss of conductivity with time and potentially to the corrosion of other parts of the fuel cell.

Although some recent examples of anhydrous proton conducting polymers utilize members of the azole family as proton solvents. Yet, still, comparatively few azoles have been studied to date due in part to the synthetic limitations of available techniques. Also, there is a problem of leaking and other forms of solvent loss which could, however, be overcome by tethering the solvent providing proton transport (e.g. imidazole, phosphoric acid) directly to a polymer backbone. An outcome is that the connectivity and percolation of solvating species is key for conduction in materials where their long-range diffusion is restricted. The low anhydrous conductivities of poly(styrenesulfonic acid) and poly(vinylphosphonic acid) for example, illustrate this effect. **Figure 10** shows azole-based materials with a polymer-bound proton solvent that exhibit measurable proton conductivity [42].

Organic photovoltaic (OPV) systems, in particular, polymer solar cells, made by solution-processed organic materials, have shown great promise as a technology for affordable electricity. Until recently, the commercialization of OPV has been hampered by the difficulty of converting laboratory-produced cell into reliable industrial-scale product performances. Unfortunately, a significant barrier to the introduction of organics into these areas has to do with inferior electrical properties as compared to traditional inorganic semiconductors.

The significant concerns by using some azole polymers arise from the azole synthesis and processing conditions. These azole polymers require very high temperatures (more than 200°C) in highly acidic media (poly(phosphoric acid)) to be synthesized and maintain its solubility after cooling [43]. Molding these polymeric materials into films also necessitates that the materials are dissolved in highly acidic media (conc. Sulfuric acid, aluminum chloride in nitromethane) [43]. Once formed, it is also challenging to get rid of trace acid, which leads to unintentional doping. The performance of organic electronic devices is hampered by the presence of acidic impurities by the interference with charge and exciton transport.

## 7. Opportunities for utilizing azoles in energy development

Fuel cells, as zero pollution systems have the potential to become alternative clean and efficient energy conversion systems [1]. From previous experiences of utilizing polybenzobisazoles in high-performance applications, all of the necessary monomers for fuel cells can now be synthesized on an industrial scale, and purification effected without the use of column chromatography. This is beneficial for large-scale synthesis of azoles. Furthermore, the ring system of benzobisazole is electron-deficient and planar, and as a consequence, leads to resilient intermolecular interactions and good charge transport properties within polymer films [44]. The key to conductivity in systems where the proton solvent is tethered to a backbone is rotational and translational mobility of the tethered moiety. While imidazole, phosphoric acid, and sulfuric acid all conduct well as mobile small-molecule liquids, the proximity to poly(vinylphosphonic acid) and stiffness of the backbone in the poly(vinylphosphonic acid) significantly limits these acids' ability to form conductive networks. By contrast, the low glass transition temperature (Tg) of the backbone and flexible tether of the imidazole-containing polymer illustrated in **Figure 10** enables proton conductivity in this material.

Azole-containing polymers are particularly attractive as anhydrous proton conductors due to their chemical diversity and comparatively small changes in physical properties in the presence of water. Sulfonic acids form stiff, immobile clusters at low hydration and absorb significant quantities of water from the atmosphere at high humidity. By contrast, less acidic moieties, such as carboxylic acids (e.g. Surlyn®, a copolymer of ethylene and methacrylic acid) hydrate much less strongly, mitigating changes in properties with changing RH [36]. The weak hydrogen bonding observed in imidazole, and by extension, the azoles in general likewise enables polymers containing these moieties to remain fluid under anhydrous conditions. This property plagues the anhydrous behaviour of sulfonated materials.

The goals of OSC technological development are not necessarily to exceed the performance of inorganics. There is a great opportunity in using combinations of the organics and inorganics. OSCs offer new device functionalities (optical transparency, chemical response, lightweight) as well as a way to produce electronic materials at a lower cost [45]. A critical factor in achieving excellent performance is to develop OPV materials (buffer materials, polymer donors, acceptors, electrodes materials and encapsulants) exhibiting the required technical and economic characteristics to be conveniently used in an industrial environment. The improvement in new materials development remains an important area of research despite the fact that CP-based OPVs are rapidly approaching the 10% power conversion efficiency recommended for them to be of commercial importance. Particularly, the advancement of effective donor materials that takes into consideration practical aspects of commercialization such as enhanced environmental and thermal stability of the resulting material, facile synthesis and purification of monomers, is still pertinent. The most attractive part about the use of OSCs instead of using traditional inorganic semiconductors, is derived from the synthetic range intrinsic in organic molecules. There exist many different ways to alter the properties of OSCs by chemical synthesis, making the OSCs easily tunable to fit the needs of a device [46].

Still needed for commercialization of azole-based materials for energy applications is the continued improvements in efficiency, stability, and cost. Ongoing research and development of azole materials, devices, and systems are making significant advances, benefiting from strong synergies with current research efforts in photovoltaics, nanotechnologies, and azole materials. In this vein, efficiencies are being improved through enhanced sunlight absorption and better surface catalysis.

## 8. Wood preservatives

Wood preservatives are known to be chemicals impregnated into wood to help with the resistance of attack by mold, decay fungi, and termites. When a wood may be in contact with humans or will be painted, waterborne wood preservatives are commonly used in their treatment. Different formulations of waterborne preservatives have been made, but only a few of these have been used commercially. Most commercial treatments contain copper ions, which give treated wood its characteristic greenish-brown colouration.

Alkaline copper quaternary (ACQ) amine, a wood preservative, is composed of 67% copper oxide and a 33% didecyldimethylammonium chloride or carbonate (a quaternary ammonium compound) [47]. Since its initial commercialization, the quaternary ammonium compound has been produced using a chloride formulation, which was later replaced with a formulation from a carbonate. Several preparations of ACQ have been commercialized, and it can be treated with an amine or ammonia carrier. Copper azole preservatives (denoted as CA-B and CA-C under American Wood Protection Association/AWPA standards) are composed of 96% ammine copper and a 4% azole. In copper azole type B, the azole is entirely composed of tebuconazole. In type C, the azole is a 50/50 mixture of propiconazole and tebuconazole. While copper azole contains a higher percentage of copper than does ACQ, the retention required for aboveground use [47] is lower and, therefore, the total amount of copper in the treated wood is less.

## 9. Azole-based materials for wood preservation

Tebuconazole had been first identified by Grundlinger and Exner (1990) as an unleachable, light and heat-stable organic biocide that provides protection against copper tolerant fungi" [48]. Kugler et al. reported of tebuconazole and propiconazole as complementing each other in terms of their efficacy against the brown rot basiodiomycete fungi [49]. The test method for assessing the performance against basiodiomycete decay fungi is EN113, and the toxic values for *Coniophoraputeana* is between 0.08–0.13 kg/$m^3$ active ingredient [48].

Copper azole type B (CA-B) is formulated from tebuconazole (4%) and amine copper (96%). In copper azole type C (CA-C), half of the tebuconazole in copper azole type B is replaced with the azole, propiconazole. Thus, the copper azole type C consists of 2% tebuconazole, 2% propiconazole and 96% amine copper. A previous preparation, copper azole type A (CBA-A) contained boric acid as a main ingredient in addition to the tebuconazole. Although copper azole is an amine formulation, it may also be formulated with an amine-ammonia compound. The ammonia may be added if the copper azole formulation is used to treat refractory wood species.

Through laboratory screening tests and extensive field trials, copper and triazole were identified as the active ingredients which could offer a viable alternative to CCA. For example, the main active ingredients in the commercial preservative TanalithR E are copper carbonate, tebuconazole and propiconazole, [50]. The ratio of actives as presented by Enviros [51] in their treated timber classification report on a percentage weight/weight basis in the preservative was copper carbonate 20%, propiconazole 0.2% and tebuconazole at 0.2%.

Azole molecules and their derivatives are among the organic corrosion inhibitors for copper that are frequently used. In this vein, density functional theory (DFT) calculations have been performed on the adsorption of four azole molecules; imidazole, 1,2,3-triazole, tetrazole, and pentazole on Cu (III) and Al (III) surfaces, and these have been characterized. It was found out that the molecules adsorb in an upright geometry onto the top site of Cu (III) only weakly, via single nitrogen atom. The chemical bonding with two nitrogen atoms to a bridge site becomes slightly preferred in all the molecules except for triazole. Molecular electronic structure is only weakly perturbed when adsorbed, and hybridization between molecular σ orbitals and metal states constitutes the molecule-surface interaction. Yet, the significant contribution to bonding comes from the electrostatic dipole interactions due to the dipole-dipole moment of azole molecules. Also, the lateral intermolecular repulsion can be significant and very long-ranged. The molecular electronegativity and chemical hardness increase linearly with increasing number of nitrogen atoms in the azole ring. The harder the molecule the more difficult the hybridization with metal states. This explains why with the increasing number of nitrogen atoms in azole ring the molecule-Cu(III) bond strength decreases linearly as: imidazole > 1,2,3–triazole > tetrazole > pentazole [52].

## 10. Current trends and future prospects of azoles in wood preservation

Wood-degrading organisms, in conditions that support their growth are generally responsible for the deterioration of many commonly-used wood species if exposed. Wood products are therefore, protected by utilizing chemical preservatives for protection against attack by decay fungi, harmful insects, or marine borers. Treating wood materials with preservatives increases their lifespan, which leads to reduced replacement costs and ensures greater resourceful use of forest trees. The extent of wood protection is dependent on the type of chemical preservative used, and the treatment method used. In terms of effectiveness, some of the chemical preservatives are better than others, while some are also more adaptable to specific applications. For long-term effectiveness, chemical preservative, and treatment method for each wood species are needed for adequate penetration and retention [53].

Wood preservatives must meet two broad criteria which include the provision of the desired wood protection in the intended end-use, and doing so without presenting unreasonable risks to people and the environment. For several decades now, copper-based wood preservatives have dominated the industrial preservation of wood for exterior applications [54].

Materials and products from nanotechnology are increasingly being produced and used for the potential they hold to provide great interests to society. As such, although still emerging, nanotechnology has been identified as a key enabling technology. One of its important areas of application is biocide preservatives for wood protection. One prominent example is copper azole, used in wood preservation through impregnations. The use of nanoscale Cu instead of bulk Cu improves the durability of wood against microbial and fungal activity due to mainly decreased viscosity of formulations and increased effective surface area of Cu, enhancing dispersion stability. These properties contribute to easier impregnation and deeper and more homogeneous uptake of reactive biocide into the wood, which allows continuous and effective protection over its lifetime. In these preservatives, copper is the main biocide, and the azole is a co-biocide. These preservatives have gained a significant market importance in the wood industry.

Copper azole which is a water-based preservative and dependent principally on copper solubilized in ethanolamine and an organic trizaole co-biocide, is a recent development. The first copper azole preservative that was developed consisted of 49% copper, 49% boric acid, and 2% tebuconazole. More recently, a copper azole preservative containing 96% copper and 4% tebuconazole has been manufactured. The copper in copper azole systems provides the primary fungicide and insecticide activity. The azole component protects against copper tolerant fungi, and thus acts as a co-biocide.

During the preservation process, proper handling and conditioning of the wood after treatment helps minimize leeching and potential environmental impacts for these preservatives. Amine keeps copper soluble in these treatment solutions. After preservative treatment, wood has to be thoroughly dried and suitably stabilized.

In the copper azole (CA–B) preservative, copper stabilization is very rapid occurring within 24 h at a retention of 1.7 kg $m^{-3}$. However, the stabilization process slows down to a large extent at higher retentions unless the wood is heated to enhance the stabilization [53].

With increasing demand for wood products, utilization of wood composites will increase, and these composite products also need to be protected with suitable wood preservatives. Thus, in the wood preservation industry, there is the need for superior alternative technologies to the traditional preservatives and pressure-treatment processes. In this instance, the development of effective and economical 3rd generation organic preservatives for wood used in areas with high or severe decay and deterioration hazards, will be interesting. The wood preservation industry needs to develop high-value products with desirable and dependable properties that have a high economic return sufficient to encourage companies to undertake the long-term and expensive research necessary to create azole-based preservatives for the future.

## 11. Challenges of azoles in wood preservation

There are known risks to aquatic communities associated with the use of azole-based wood preservatives. For instance, it has been established that micronized copper azole represents a source of harm to marine benthic communities comparable to that from copper salts, such as copper sulfate. There is therefore a need for better understanding of benthic community interactions when exposed to nanomaterial stress [55].

The increased use of copper azole as wood preservative for residential construction has exposed the preservative as causing corrosion to fasteners. There is limited evidence on the effects of these preservatives on the corrosion rate of the fasteners. However, Simpson Strong Tie has a technical bulletin publication which indicates that both ACQ and copper azole are roughly twice as corrosive as chromate copper arsenate (CCA) and gives recommendations on fastener types for a given environment and preservative [56].

Recently, however, the durability of fasteners in preservative-treated wood has been a key concern. Changes in legislation and certification in some countries have restricted the use of chromate copper arsenate (CCA), which used to be the most extensively used waterborne wood preservative [57]. Ensuing these changes, several different wood preservatives have come to the market, some of which are much more corrosive than CCA [58].

Prospective health effects of exposure to copper azole preservative are shown in **Table 2**. Tebuconazole is slightly persistent in the environment, and it is not mobile. Also, light intensely increases the degradation progression. Tebuconazole degradation is approximately 20% in water according to the Organization for Economic Co-operation and Development's Test Guideline 301C. Its half-life in soil is around 100 days. Tebuconazole is considered moderately toxic to aquatic organisms and has a slight potential to bioconcentrate, but it is rapidly eliminated from fish [59].

## 12. Opportunities for utilizing azoles in wood preservatives

Both old and new structures are susceptible to mold infestation in the absence of moisture. Treatment of wood products with nontoxic, nonvolatile fungicides adds a level of protection against mold infestation. However, these preservatives have corrosion problems. Use of azole-based fungicides to protect wood from indoor mold infestations is one strategy to address this problem. Although a lot of recent

| *(Wolman® NB)* | | **Possible health effects** | |
|---|---|---|---|
| **Exposure category (Route of Entry)** | **Type of exposure** | **Short-term exposure** | **Long-term exposure** |
| Estimated daily intake from various sources (air, water, food) with limited to no heath effect[a] | | | |
| Copper (an essential element) | 2.47 mg/day | | |
| Eye contact[a,b] | Direct contact | CA-B concentrate is corrosive | Ulceration, may cause irreversible damage |
| | | Will cause irritation, pain and reddening | |
| Skin contact[a,b] | Significant skin contact with concentrates | Short term (up to 1 hour) | Long term |
| | | Mild to moderate skin irritation, inflammation, reddening | Severe irritation, ulceration, chemical burns |
| ACGIH threshold limit value-time weighted averages (TWAs)[c] | Ethanolamine: 8 $mg/m^3$ air 3 ppm | | |
| Exposure to airbome contaminant or dust inhalation[a,b] | Inhalation of mists, droplets or dust of concentrates | May cause upper respiratory tract irritation | Moderate to severe irritations of mucous membrane, nose, throat and lungs |
| | | Moderate irritation of nose, throat and lungs | |
| ACGIH threshold limit value-time weighted averages (TWAs)[c] | Copper (dusts and mists): 1.0 mg Cu/ $m^3$ air | Irritation of eyes | Irritation of eyes |

[a]*International Labour Organization ICSC Card database, http://www.ilo.org/dyn/icsc/showcard.home*
[b]*Agency for Toxic Substances and Disease Registry (ATSDR) http://www.atsdr.cdc.gov/substances/index.asp*
[c]*American Conference of Governmental Industrial Hygienists (ACGIH): http://www.acgih.org/tlv/*

**Table 2.**
*Prospective health effects of exposure to copper azole preservative [59].*

research has been conducted in this area, no attempt has been made to summarize all the recent advances, and confusion exists about the corrosiveness of alternatives to CCA and proper materials selection for use in treated wood. Thus, opportunities exist in searching for appropriate materials for the requisite preservatives.

Copper azole is a major copper-based wood preservative that has come into wide use in both developed and developing countries following restraints on CCA. The use of copper azole and other preservatives are directed by national and international specifications, which give the requirement for the volume of preservative application for a specific wood end-use. In terms of chemical composition, copper azole is similar to ACQ. The difference between them, however, is the dissolved copper preservative which is augmented by an azole co-biocide like organic triazoles such as tebuconazole or propioconazole, in the copper azole preservative. These preservatives are also used in food crops protection, instead of the quaternary biocide which is used in ACQ [60]. The azole co-biocide produces a copper azole compound that is effective at lower retentions than required for equivalent ACQ performance.

Wood treated with copper azole is marketed widely across many international markets. The AWPA standard retention for CA-B is 0.10 $lb/ft^3$ for above ground applications and 0.21 $lb/ft^3$ for ground contact applications. Copper azole type C has been presented under the Wolmanized and Preserve brands. The AWPA standard retention for CA-C is 0.06 $lb/ft^3$ for above ground applications and 0.15 lb/ $ft^3$ for ground contact applications. Opportunities exist for local standardization of these copper-azole preservatives as well as expanded local markets. Also, research question on how azoles contribute to the leaching of copper in treated wood is significant.

## 13. Environmental impact of azoles utilization

Azoles are widely used and efficient fungicides commonly employed to treat and prevent fungal diseases in humans and animals, as well as in food production, horticulture and wood industry. Residues of azoles in nature are regarded as environmental toxins and suggested to have general endocrine-disrupting properties. It has been suggested that triazole resistance has evolved in the environment and could be driven by the selective pressure of azole fungicides [61].

A significant challenge facing treated wood products is the lack of an effective strategy for handling treated wood that has been removed from service [61]. Until recently, the fixation processes of the amine wood preservatives were poorly understood, but ongoing research in North American university laboratories is beginning to expand the knowledge base considerably [62].

Some research works have shown that copper azole-treated wood can be chipped or flaked and recycled to form durable panel products or wood composites. However, this type of recycling has not gained significant commercial acceptance because of concerns with processing the treated wood. Recycling of the treated wood releases the preservatives into the panel fabrication process, which leeches into the environment, with consequent adverse impacts [63].

The widespread use of azoles in biomass preservation can affect the environment and the phytopathogens therein, with concomitant medical implications [64]. Accordingly, with azole treatment, fungi causing important human mycoses may develop azole-resistance [64, 65]. Azole as fungicide is very significant, as some human diseases are caused by fungi such as Aspergillus, Histoplasma, Coccidioides and Cryptococcus that survive in different environments [65].

## 14. Conclusion

Azole-based solar cells, fuel cells, and wood preservatives are of critical importance in energy applications and wood treatment, respectively. Multiple benefits are accrued from exploring this type of research in organic chemistry. Firstly, it will help in unraveling the intrinsic chemical behavior of azoles and their interactions with other molecules. Secondly, it will significantly help in the advancement of novel synthetic methods. Thirdly, using spectral methods for the characterization of a set of compounds could create benchmarks for similar molecules. In the next step, structure–activity relationships of azoles applications in energy and wood preservation will enhance their utilization. Finally, biological evaluation of the synthesized azole compounds may explore lead-compounds for further structural fine-tuning. Among the broad array of organic compounds, those incorporating one or more sulfur or nitrogen atoms like azoles are of great significance because of their unique properties imparted by these elements.

There are four main determinants as to how well an OSC will perform in a device: oxidative and thermal stability, properly aligned energy levels, good thin-film morphological characteristics, and purity and defects. The more researchers explore azole functions, the greater the possibility for OPV to demonstrate at last its enormous potential on the industrial scale.

Recommendations for additional steps to assess the risks and consequences of the environmental usage of azole derivatives are pertinent. With azoles applications in both solar cells and fuel cells, where electronic excitations and ions mobility properties respectively of the azoles are taken advantage of, a solar PV/fuel cell hybrid energy systems for stationary applications employing azoles, could be embarked upon, and preliminary energy and exergy efficiency analyses performed for the hybrid energy system. Such a system, built on different scientific principles, can convert solar energy and chemical energy of fuel to electrical energy simultaneously within the same system.

For azole-based fuel cells and solar cells to achieve widespread production and adoption in energy applications, especially in developing countries, it is necessary to decrease their cost of production. Also, an increase in research output in the area and improvement in the range of environmental conditions under which they effectively operate will ensure widespread production and adoption in energy applications.

A broad approach that combines preservative formulations, treatment efficiency, component interaction studies, and carefully designed strategies for azole-based wood preservative utilization is needed. This approach will increase preservative efficacy, corrosion resistance, and reduce the risk of environmental pollution, and prevent azole-resistant infections. Improved research work, including azole-based preservative optimization and modeling, is a significant key to a better understanding of the magnitude of this emerging approach.

## Author details

Nana Derkyi
University of Energy and Natural Resources, Sunyani, Ghana

*Address all correspondence to: nana.derkyi@uenr.edu.gh

## References

[1] Wanga Y, Chen KS, Mishler C, Cho SC, Adroher XC. A review of polymer electrolyte membrane fuel cells: Technology, applications, and needs on fundamental research. Applied Energy. 2011;**88**:981-1007

[2] Zhang H, Shen PK. Recent development of polymer electrolyte membranes for fuel cells. Chemical Reviews. 2012;**112**:2780. DOI: 10.1021/cr200035s

[3] Wu L, Zhang Z, Ran J, Zhou D, Li C, Xu T. Advances in proton-exchange membranes for fuel cells: An overview on proton conductive channels (PCCs). Physical Chemistry Chemical Physics. 2013;**15**:4870. DOI: 10.1039/C3CP50296A

[4] Kreuer KD. Ion conducting membranes for fuel cells and other electrochemical devices. Chemistry of Materials. 2014;**26**:361. DOI: 10.1021/cm402742u

[5] Crossman M, Simm J. Manual on the Use of Timber in Coastal and River Engineering. London: Thomas Telford Publishing; 2004

[6] Borges LMS, Cragg SM, Busch S. A laboratory assay for measuring feeding and mortality of the marine wood borer Limnoria under forced feeding conditions: A basis for a standard test method. International Biodeterioration & Biodegradation. 2009;**63**:289-296

[7] Hill CAS. Wood Modification: Chemical, Thermal and Other Processes. Chichester: Wiley; 2006

[8] Eicher T, Hauptmann S. The Chemistry of Heterocyclics: Structure, Reactions, Synthesis, and Applications. 2nd ed. Weinheim, Germany: John Wiley & Sons; 2003

[9] Souza R, Miranda L. Strategies towards the synthesis of N2-substituted 1,2,3-triazoles. Anais da Academia Brasileira de Ciências. 2019;**91**:24. DOI: 10.1590/0001-3765201820180751

[10] Shawa AY, Chang C, Hsu M, Lu P, Yang C, Chen H, et al. Exploration of carbamide derived pyrimidine-thioindole conjugates as potential VEGFR-2 inhibitors with anti-angiogenesis effect. European Journal of Medicinal Chemistry. 2010;**45**:2860-2867

[11] Kuznetsov YI, Kazansky LP. Physicochemical aspects of metal protection by azoles as corrosion inhibitors. Russian Chemical Reviews. 2008;**77**(3):219-232

[12] Gupte M, Kulkarni P, Ganuli BN. Antifungal antibiotics. Applied Microbiology and Biotechnology. 2002;**58**:46-57

[13] Naarmann H. Polymers, electrically conducting. In: Ullmann's Encyclopedia of Industrial Chemistry. KGaA: Wiley-VCH Verlag GmbH & Co; 2000

[14] Hösel M, Angmo D, Krebs FC. Organic solar cells. In: Ostroverkhova O, editor. Handbook of Organic Materials for Optical and (Opto) electronic Devices: Properties and Applications. Cambridge, United Kingdom: Woodhead Publishing Series; 2013. pp. 473-507. DOI: 10.1533/9780857098764.3.473

[15] Kietzke T. Advances in OptoElectronics. Egypt: Hindawi Publishing Corporation; 2007. p. 40285

[16] Loo YL, McCulloch I. Progress and challenges in commercialization of organic electronics. MRS Bulletin. 2008;**33**(7):653-705

[17] Bernius MT, Inbasekaran M, O'Brien J, Wu W. Progress with light-emitting polymers. Advanced Materials. 2000;**12**(23):1737-1750

[18] Woolley DW. Some biological effects produced by benzimidazole and their reversal by purines. The Journal of Biological Chemistry. 1944;**152**:225-232

[19] Sheehan DJ, Hitchcock CA, Sibley CM. Current and emerging azole antifungal agents. Clinical Microbiology Reviews. 1999;**12**:40-79

[20] Fromtling R. Overview of medically important antifungal azole derivatives. Clinical Microbiology Reviews. 1988;**1**:187-217

[21] Sanghi S, Tuominen M, Coughlin EB. Hybrid inorganic–organic proton-exchange membranes containing 1H-1,2,3-triazole moieties. Solid State Ionics. 2010;**181**:1183. DOI: 10.1016/j.ssi.2010.06.051

[22] Je-Deok K, Toshiyuki M, Shigenobu H, Itaru H. Anhydrous proton-conducting properties of Nafion–1,2,4-triazole and Nafion–benzimidazole membranes for polymer electrolyte fuel cells. Journal of the Electrochemical Society. 2007;**154**:4. DOI: 10.1149/1.2436611

[23] Song MK, Zhu X, Liu M. A triazole-based polymer electrolyte membrane for fuel cells operated in a wide temperature range (25-150 °C) with little humidification. Journal of Power Sources. 2013;**241**:219

[24] Fátima C, Teixeira C, Rangel M. Synthesis of azole phosphonates precursors for proton-exchange membrane for application in high temperature PEM fuel cells. In: Proceedings of the 4th International Seminar on Advances in Hydrogen Energy Technologies: Oportunities and Challenges in a Hydrogen Economy. Viana do Castelo – Portugal; November 10-11, 2011

[25] Wolfe JF. Encyclopedia of Polymer Science and Engineering. Vol. 11. New York, NY: John Wiley and Sons; 1988. p. 601

[26] Osaka I, Takimiya K, McCullough RD. Benzobisthiazole-based semiconducting copolymers showing excellent environmental stability in high-humidity air. Advanced Materials. 2010;**22**:4993. DOI: 10.1002/adma.201002117

[27] Laberty-Robert C, Valle K, Pereira F, Sanchez C. Design and properties of functional hybrid organic–inorganic membranes for fuel cells. Chemical Society Reviews. 2011;**40**:961-1005

[28] Li A, Cao Z, Li Y, Yan T, Shen P. Structure and dynamics of proton transfer in liquid imidazole. A Molecular Dynamics Simulation. The Journal of Physical Chemistry. B. 2012;**116**:12793-12800

[29] Schuster MFH, Meyer WH, Kreuer KD. Toward a new type of anhydrous organic proton conductor based on immobilized imidazole. Chemistry of Materials. 2004;**16**:329-337

[30] Edmondson C, Fontanella J, Chung SH, Greenbaum SG, Wnek G. Complex impedance studies of S-SEBS block polymer proton-conducting membranes. Electrochimica Acta. 2001;**46**(10):1623-1628. DOI: 10.1016/S0013-4686(00)00762-3

[31] Gustian I, Çelik SÜ, Bozkurt A. Novel proton conductive hybrid membranes based on sulfonated polysulfone and benzotriazole. Journal of Materials Research. 2012;**27**:2650-2656

[32] Sen U, Unugurcelik S, Ata A, Bozkurt A. Anhydrous proton conducting membranes for PEM fuel cells based on nafion/azole composites. International Journal of Hydrogen Energy. 2008;**33**:2808-2815

[33] Balzani V, Credi A, Venturi M. Photochemical conversion of solar energy. ChemSusChem. 2008;**1**:26-58

[34] Wöhrle D, Meissner D. Organic solar cells. Advanced Materials. 1991;**3**:129-138

[35] Tang CW. Two-layer organic photovoltaic cell. Applied Physics Letters. 1986;**48**(2):183-185

[36] Facchetti A. $\pi$-Conjugated polymers for organic electronics and photovoltaic cell applications. Chemistry of Materials. 2011;**23**:733-758. DOI: 10.1021/cm102419z

[37] van Mullekom HAM, Vekemans JAJM, Havinga EE, Meijer W. Developments in the chemistry and band gap engineering of donor-acceptor substituted conjugated polymers. Materials Science and Engineering R. 2001;**32**(1):1-40. DOI: 10.1016/S0927-796X(00)00029-2

[38] Sariciftci NS. Plastic photovoltaic devices. Materials Today. 2004;**7**(9):36-40

[39] Nakamura T, Okazaki S, Arakawa N, Satou M, Endo M, Murata Y, et al. Synthesis of azole-fused benzothiadiazoles as key units for functional $\pi$-conjugated compounds. Journal of Photopolymer Science and Technology. 2017;**30**(5):561-568. DOI: 10.2494/photopolymer.30.561

[40] Zhou E, Nakamura M, Nishizawa T, Zhang Y, Wei Q, Tajima K, et al. Synthesis and photovoltaic properties of a novel low band gap polymer based on N-substituted dithieno [3,2-b:2′,3′-d]pyrrole. Macromolecules. 2008;**41**(22):8302-8305

[41] Blouin N, Michaud A, Leclerc M. A low-bandgap poly(2,7-carbazole) derivative for use in high-performance solar cells. Advanced Materials. 2007;**19**(17):2295-2300

[42] Scharfenberger G, Meyer WH, Wegner G, Schuster M, Kreuer KD, Maier J. Anhydrous polymeric proton conductors based on imidazole functionalized polysiloxane. Fuel Cells. 2006;**6**:237-250

[43] Hu XD, Jenkins SE, Min BG, Polk MB, Kumar S. Rigid-rod polymers: Synthesis, processing, simulation, structure, and properties. Macromolecular Materials and Engineering. 2003;**288**(11):823-843

[44] Conti F, Willbold S, Mammi S, Korte C, Lehnert W, Stolten D. Carbon NMR investigation of the polybenzimidazole–dimethylacetamide interactions in membranes for fuel cells. New Journal of Chemistry. 2013;**37**:152

[45] Start PR, Mauritz KA. Surlyn®/silicate nanocomposite materials via a polymer in situ sol–gel process: Morphology. Journal of Polymer Science Part B: Polymer Physics. 2003;**41**:1563-1571

[46] Celik SU, Bozkurt A, Hosseini SS. Alternatives toward proton conductive anhydrous membranes for fuel cells: Heterocyclic protogenic solvents comprising polymer electrolytes. Progress in Polymer Science. 2012;**3**:1265

[47] Anon A. WPA U1-06: Use Category System. Granbury, TX: American Wood Preservers' Association; 2007

[48] Grundlinger R, Exner O. Tebuconazole, a new wood preserving fungicide. In: Proceedings of the 21st Annual Meeting of IRG/WP/3634. Rotorua, New Zealand; 1990. p. 3

[49] Kugler M, Bruns R, Jaetsch T, Spetmann P. Fungicidal Mixtures for Wood Preservation. Pittsburg, US: Lanxess Corp; 2008

[50] Arch Timber Protection. Best Practice - Treatment Plant Design

and Site Maintenance. Technical, Environmental & regulatory Affairs, Ref Gl/08; 2008

[51] Enviros Consulting Ltd. Assessment of File on Classification of Treated Timber Waste. Final Report prepared for the Environmental Protection Agency. Manchester; 2003. pp. 8-10

[52] Kovačević N, Kokalj A. DFT study of interaction of azoles with Cu(111) and Al(III) surfaces: Role of azole nitrogen atoms and dipole–dipole interactions. Journal of Physical Chemistry C. 2011;**115**(49):24189-24197. DOI: 10.1021/jp207076w

[53] Stan T. Lebow Wood Preservation General Technical Report, Vol. 15. FPL–GTR–190; 2010. pp. 1-28

[54] Joerg H. Copper-based wood preservative formulations. In: Proceedings of the Workshop on Lifecycle Impacts of Copper Nanomaterials Released from Timber Preserving Impregnations. Venice, Italy: University Ca' Foscari; January 2016

[55] Kay TH, Lisa P, Anthony AC, Marguerite P, Mark C, David K, et al. Effects of micronized and nano-copper azole on marine benthic communities. Environmental Toxicology and Chemistry. 2017;**37**:362-375. DOI: 10.1002/etc.3954

[56] Simpson Strong Tie. Corrosion Fact Sheet. Simpson Strong Tie Technical Bulletin T-PTWOOD06. 2006. Available from: www.strongtie.com/ftp/bulletins/T-PTWOOD06.pdf

[57] Lebow S. Wood preservation. In: Wood handbook – Wood as an engineering material. Madison, WI: U.S. Department of Agriculture, Forest Service, Forest Products Laboratory; 2010. p. 508

[58] Kear G, Wu HZ, Jones MS. Weight loss studies of fastener materials corrosion in contact with timbers treated with copper azole and alkaline copper quaternary compounds. Corrosion Science. 2009;**51**:252-262

[59] Occupational Safety & Health Administration (OSHA). Permissible Exposure Limits (PELs). Available from: http://www.osha.gov/dsg/topics/pel/index.html

[60] Lebow S. Alternatives to Chromated Copper Arsenate for Residential Construction. US Department of Agriculture. Report Research Paper FPL-RP-618; 2004

[61] Connell M. Waste management options in the wood protection industry. Journal of the Institute of Wood Science. 1999;**15**(1):51-58

[62] JCF W, editor. Primary Wood Processing: Principles and Practice. 2nd ed. Berlin, Heidelberg: Springer; 2006. p. 336

[63] Kartal SN, Clausen CA. Leachability and decay resistance of particleboard made from acid extracted and bioremediated CCA-treated wood. International Biodeterioration & Biodegradation. 2001;**47**:18391

[64] Hof H. Critical annotations to the use of azole antifungals for plant protection. Antimicrobial Agents and Chemotherapy. 2001;**45**:2987-2990

[65] Azevedo MM, Faria-Ramos I, Cruz LC, Pina-Vaz C, Rodrigues AG. Genesis of azole antifungal resistance from agriculture to clinical settings. Journal of Agricultural and Food Chemistry. 2015;**63**:7463-7468